Buffalo Bill's
Wild West

Scottsdale Gets Bill's Best

Selections from the
Buffalo Bill Historical Center
Cody, Wyoming

January 19 – April 28, 2002

Fleischer Museum

Buffalo Bill's Wild West

Scottsdale Gets Bill's Best

Selections from the Buffalo Bill Historical Center

Front cover:
Portrait
Strobridge Lithographic Co.
Color lithograph poster; 40" x 58"
Cincinnati, Ohio, 1908
Mort and Donna Fleischer Collection

Back cover:
"The Farewell Shot" Positively the last appearance of Col. W.F. Cody (in the saddle), "Buffalo Bill."
U.S. Lithograph Co.
Color lithograph poster, Russell-Morgan, Print; 28" x 41"
Cincinnati and New York, c. 1910
Buffalo Bill Historical Center, Cody, WY; 1.69.137

Page II & III:
Custer's Last Stand
Edgar S. Paxson (1852–1919)
1899, o/c, 70.5" x 106"
Buffalo Bill Historical Center, Cody, WY; 19.69

Page VI:
Col. William F. Cody (detail)
Rosa Bonheur (1822–1899)
1889, o/c, 18.5" x 15.25"
Buffalo Bill Historical Center, Cody, WY
Given in memory of William R. Coe and Mai Rogers Coe; 8.66

17207 North Perimeter Drive
Scottsdale, Arizona 85255
480.585.3108 www.fleischer.org
A city supported facility

Scottsdale, Arizona 85255
ISBN #0-9617882-9-1

Edited by Robert B. Pickering, Ph.D.
Designed by Fleischer Museum
Printed by Image Communications

Fleischer Museum gratefully acknowledges the generous support of:

AJ's Purveyors of Fine Foods
Charleston's Restaurants
Desert Ridge Marketplace
GE Capital Franchise Finance
Hampton Inn & Suites
Hyatt Regency Scottsdale at Gainey Ranch
Image Communications
Kutak Rock
P.F. Chang's China Bistro
Rawhide Wild West Town
Scottsdale Charros
Scottsdale Convention & Visitors Bureau
Scottsdale Western Art Association
Wells Fargo Bank
Z Tejas

Buffalo Bill's Wild West

Scottsdale Gets Bill's Best

Selections from the Buffalo Bill Historical Center

This catalogue was published in conjunction with the exhibition
at Fleischer Museum
January 19 – April 28, 2002

Lewis and Clark and Sacagawea
Henry Lion (1900–1966)
c. 1963, bronze, 35.25" x 23.75" x 28.25"
cast no. 2, Roman Bronze Works Inc., NY
Buffalo Bill Historical Center, Cody, WY
Gift of Charles S. Jones; 27.64

This sculpture commemorates the exploration of the territory acquired by the United States government through the Louisiana Purchase. The leaders of the exploring party, Meriwether Lewis and William Clark, are aided by the Shoshone woman Sacagawea, whose presence confers an acceptance of the goals of the exploring party.

Acknowledgement
by Donna H. Fleischer, Executive Director, Fleischer Museum

The City of Scottsdale has been called the "West's Most Western Town," and January 19, 2002 marks the opening of an exhibition collaboration between this western town, Fleischer Museum, and the premier western heritage museum, Buffalo Bill Historical Center. Fleischer Museum will showcase over 100 of the best images and icons of the myth and reality of the great American West. The West is the essence of American soul and has always symbolized the freedom and the spirit of rugged individualism. From craggy peaks to lush valleys and flowing rivers, the West represented the homeland for numerous Indian Nations and promise for waves of immigrants from all over the world.

Lewis and Clark, Buffalo Bill, General George Custer, Sitting Bull, and White Man Runs Him are well-known historical personages. In addition, there are those whose lives created the truths and the legends of the West although their names have been lost in time. They were men and women who lived, fought, died, and ultimately shaped the West that we inherited. These events characterized the emerging country and the American psyche. *Buffalo Bill's Wild West, Scottsdale Gets Bill's Best* espouses the spirit of the West, its frontiers and challenges, and exemplifies the landscapes, the people, the events and the stories that are the West.

Fleischer Museum almost closed this year after the sale of Franchise Finance Corporation of America, which houses the Museum, had it not been for the City of Scottsdale. The City recognized the cultural and economic contribution the Museum has made to the area in the last eleven years and has agreed to build a new Museum in Scottsdale. We thank the Scottsdale Tourism Development Commission Board, Scottsdale Convention and Visitors Bureau, and the citizens of Scottsdale for their enthusiastic support, as well as Mayor, Mary Manross; City Manager, Jan Dolan; and Council members: Cynthia Lukas, Ned O'Hearn, David Ortega, Robert Pettycrew, Tom Silverman, and George Zraket.

We also recognize and thank Wyoming Senator and Chairman of the Board of the Buffalo Bill Historical Center (BBHC), Alan Simpson for his encouragment along with the following BBHC personnel for their time and diligent expertise in assembling the information and materials to expedite the exhibition: former Executive Director, Byron Price; Wally Reber, Associate Director; Robert B. Pickering, Ph.D., Deputy Director for Collections and Education; Nathan Bender, Housel Curator, McCracken Research Library; Sarah E. Boehme, Ph.D., The John S. Bugas Curator, Whitney Gallery of Western Art; Julie Coleman, Whitney Curatorial Assistant; Ann Marie Donoghue,

Associate Registrar; Emma I. Hansen, Curator, Plains Indian Museum; Marguerite House, Executive Assistant; Jan Jones, Communications Specialist; Warren Newman, Curatorial Assistant, Cody Firearms Museum; Gina Schneider, Assistant to the Director; Dean Swift, Gift shop Manager; Rebecca S. West, Plains Indian Museum Curatorial Assistant.

Sponsors are alway an important aspect of Museum exhibitions and we greatly appreciate the enthusiastic support of Paul Bossidy, President and CEO of CEF GE Capital and Bruce Nelson, President GE Capital Franchise Finance for allowing us to fulfill the Museum's expectations. Also Christopher Volk, COO GE Capital Franchise Finance and Steve Schwanz, Sr. Vice President Southwest region.

Other sponsors who have been vital to the success of this event are: AJ's Purveyors of Fine Foods; Charleston's Restaurants; Desert Ridge Marketplace; Hampton Inn & Suites; Hyatt Regency Scottsdale at Gainey Ranch; Image Communications; Kutak Rock; P.F. Chang's China Bistro; Rawhide Wild West Town; Scottsdale Charros; Scottsdale Convention & Visitors Bureau; Scottsdale Western Art Association; Wells Fargo Bank and Z Tejas.

I also extend a very special thank you to the dependable and ever malleable Fleischer staff who include Joan Hoeffel who assisted in creating the catalogue and collateral material, Karen Turek, Garth Lorenz and Oscar Ornelas.

Mission Saddle
Edward H. Bohlin, Inc.
Handmade silver medallions portraying all missions built by early Spanish settlers in California, 1930
Brown leather, floral cutout overlay back of cantle and silver horn
Buffalo Bill Historical Center, Cody, WY
Gift of H. Peter Kriendler, and his brothers Mac and Bob Kriendler;
1.69.376 A, B, C

BY SENATOR ALAN K. SIMPSON, CHAIRMAN BUFFALO BILL HISTORICAL CENTER BOARD OF TRUSTEES

Buffalo Bill learned a long time ago that it was darned important to take his show to the people. Through the cooperation of our friends, Mort and Donna Fleischer, Fleischer Museum and the City of Scottsdale, the Buffalo Bill Historical Center (BBHC) of Cody, Wyoming is continuing in ol' Bill's tradition and now sharing some of the most important objects and works of art with folks in Arizona.

The West, wild or not, embraces much more than Buffalo Bill, the man, or his *Wild West* show. He still stands as an icon of the West and also as a visionary and an entrepreneur. While enjoying this exhibit, you will see many other icons of the West and will come to appreciate the complexity and diversity of the Old West.

The Buffalo Bill Historical Center is the premier museum of the West. Through this exhibit and catalogue, we present to you not only a wealth of art and objects but also, multiple perspectives on the West. For an even more exciting view, I invite you all to visit our five museums that compose the BBHC: the Buffalo Bill Museum, the Plains Indian Museum, the Whitney Gallery of Western Art, the Cody Firearms Museum, and our newest endeavor, the Draper Museum of Natural History (opening June 4, 2002). Together, these splendid museums encompass 6.9 acres under one roof. It truly is a place where "the wild meets the West." See you in Cody!!

Poster, 1
The Enquirer Job Printin
Color lithograph poster, 110.5" x 81",
Buffalo Bill Historical Center, Cody
1.69.

Foreword

By Robert B. Pickering, Ph.D., Deputy Director for Collections and Education

The Buffalo Bill Historical Center (BBHC) is pleased to collaborate with Fleischer Museum to present *Buffalo Bill's Wild West, Scottsdale Gets Bill's Best.* Buffalo Bill knew that it was important to take the *Buffalo Bill Wild West* show to the people. By sharing many of the BBHC's treasures with the people of Scottsdale, we are following ol' Bill's legacy.

The Buffalo Bill Historical Center is the premier museum of the American West. Through our large and diverse collections, great themes of the West—the realities and the myths—are explored. The BBHC is perhaps unique as a museum. In fact, we are five museums: the Buffalo Bill Museum, the Whitney Gallery of Western Art, the Plains Indian Museum, the Cody Firearms Museum and the Draper Museum of Natural History (opening in June 2002). In addition, the McCracken Research Library is the repository of great photography, papers, and book collections that document the West. Taken together the museums of the BBHC provide an unparalleled opportunity to explore the West as it was and as it has been imagined. Recently, the Cody Institute of Western American Studies (CIWAS) was established to explore important western ideas and to create opportunities to take those discussions to people who know Cody, Wyoming, only as a small dot on a large map. Through developing new kinds of symposia, the BBHC website and other media, CIWAS will serve a diversity of people who want to know about the West but may not be able to come to Cody.

This catalogue has been organized according to four broad themes: the landscape, the people, the events, and the stories. While volumes have been written about each one of these topics, this exhibit attempts to capture the essence of each theme through some of the finest examples of art and objects from all of the collections of the BBHC. Finely designed and crafted Plains Indian clothing speaks of the significance of the animals, the land and the cosmos surrounding human beings on the Plains. Paintings and sculptures present the majesty of the great western vistas, document the events that took place, and provide touchstones for the mythologies that grew out of the West. Actual documents, guns, photos, and other personal material put the West in a human context. Sitting Bull, Buffalo Bill, Annie Oakley, and the long parade of individuals created lives for themselves and thereby have become the history and the legends of the western life.

The West is a truly American place. Some scholars define it by geography; others say it is a perspective or an idea. This exhibit does not take any particular side in that debate. However, we hope this extraordinary assemblage provokes the visitor to contemplate the West, to appreciate our heritage, and to understand that the West is still a powerful part of American culture.

Days of Long Ago
Henry Farny (1847–1916)
1903, o/b, 37.5" x 23.75"
Buffalo Bill Historical Center, Cody, WY; 6.75

The high horizon and the shadowy, sheltering trees form an intimate setting for Farny's family group even within the vastness of the western landscape. With the title *Days of Long Ago*, Farny wistfully places the scene in some distant, but not specific, time. Farny's vision of the Indian rests in tranquility.

Beauty and Grandeur: The Landscape of the American West

by Sarah E. Boehme, Ph.D., The John S. Bugas Curator, Whitney Gallery of Western Art

Expectations of "beauty and grandeur" colored the perceptions of early explorers traveling into the American West and found fulfillment on the canvases of generations of artists who followed. Samuel Seymour, the first Euro-American artist to depict the Rocky Mountains, journeyed to the West to document information within a military and scientific context; he was appointed as the artist for Major Stephen H. Long's 1819–1820 expedition. Yet Long's description of Seymour's duties indicated that aesthetic qualities governed the artist's orders, "Mr. Seymour, as painter for the expedition, will furnish sketches of landscapes, whenever we meet with any distinguished for their beauty and grandeur."[i] Seymour's abilities to portray extraordinary beauty and grandeur may have been constrained by his own rudimentary artistic training and by the expedition's rigors, but his works brought to American and European audiences a glimpse of the landscape while indicating the challenges of portraying the vistas of vast American plains and distant mountains.

Artists Albert Bierstadt and Thomas Moran eagerly embraced the challenges of portraying the western landscape and, with their talents and training, brought a new level of artistic accomplishment to the West. Their emphasis on grandeur conveyed a compelling view to their mid-to-late nineteenth century audience. In their first trips to the region, both Bierstadt and Moran accompanied government expeditions, not as official, salaried members of the parties, but rather from their own initiatives. During his 1859 trip, Bierstadt likened the Rocky Mountains to the Bernese Alps, a comparison that highlights his inclination toward majestic scenery. Always rooted in his experiences, Bierstadt heightened the emotional content of his portrayals by compressing or exaggerating landscape elements to create the most dramatic scene. Similarly, Moran depicted notable sites, such as Yellowstone National Park, but constructed his paintings to create glorious pictorial prospects not available from any single viewpoint.

At a time when the West was the site of tremendous change and upheaval, artists sometimes chose to emphasize the quiet beauty of nature. The Arapaho artist who painted a Ghost Dance shirt with stars and birds expressed a Native American concept of place within the universe.[ii] Greatly aware of modern changes such as the railroads, Henry Farny often conveyed a longing for an Eden or a Golden Age of a bucolic past. Carl Rungius captured wildlife in a moment of stillness, perceptible against the natural setting while being part of it. Photographer L.A. Huffman portrayed cattle as an integral part of a peaceful, seemingly endless, West. In contrast, artists such as Frederic Remington and Philip R. Goodwin depicted the West as a setting of danger, with natural elements as a threat to human existence within the landscape.

Renewing the sense of awe and wonder evoked by the western landscape is one challenge for the contemporary artist, who works within a context of a surfeit of visual images. Wilson Hurley reinvigorates the panoramic view with the knowledge available from the aerial perspectives of airplanes, Fritz Scholder abstracts the landscape with an emotional use of color and brushstroke, and Bill Schenck playfully recalls the imagined West. Multiple interpretations result, but a faith in the power of the landscape prevails. New audiences approach the West still expecting to find scenes of beauty and grandeur.

Edwin James, comp., *Account of an Expedition from Pittsburgh to the Rocky Mountains Performed in the Years 1819 and 1820, by Order of the Honorable J.C. Calhoun, Secretary of War, under the Command of Major Stephen H. Long of the U.S. Top. Engineers,* 3 vols. (London: Longman, Hurst, Rees, Orme and Brown, 1823) I: 3.

Cross-cultural comparisons are more complex than can be effectively dealt with in this brief essay. See elsewhere in this publication for a more multifaceted exploration of Native points of view.

Winchester Model 1907 Deluxe Semi-Automatic Rifle
Maker: Winchester Repeating Arms Co., New Haven, CT
Serial Number: 16,417
Caliber: .351 (.351 Winchester Self-Loading centerfire cartridge)
Date: 1909
Buffalo Bill Historical Center, Cody, WY
Gift of Olin Corporation, Winchester Arms Collection; 1988.8.1

John A. Gough engraved this rifle with deer on both right and left sides of its receiver. This rifle was exhibited by Winchester at the Panama-Pacific 1915 Exposition.

Distant View of the Rocky Mountains
Samuel Seymour, artist (c. 1775–c. 1823)
I. Clark, engraver
drawn 1820, engraving printed 1823 in Edwin James, *Account of an Expedition from Pittsburgh to the Rocky Mountains Performed in the Years 1819 and 1820, by Order of the Honorable J.C. Calhoun, Secretary of War, under the Command of Major Stephen H. Long of the U.S. Top. Engineers, vol. 1* (London: Longman, Hurst, Rees, Orme and Brown, 1823)
Buffalo Bill Historical Center, Cody, WY

Seymour painted the first landscapes of the Rocky Mountains in the Euro-American tradition of perspectival views. He served as the official artist appointed to Major Stephen H. Long's expedition of 1819–1820.

Last of Their Race
John Mix Stanley (1814–1872)
1857, o/c, 43" x 60"
Buffalo Bill Historical Center, Cody, WY; 5.75

An allegory on the theme of the Indian as a dying race, this painting depicts remnants of the tribes pushed to the edge of the ocean with the sun setting in the distance and buffalo skulls forecasting the end. Stanley arranged his representatives of the tribes and ages in a pyramid, giving a classical composition to his painting.

A Herd of Bison Crossing the Missouri River
William Jacob Hays, Sr. (1830–1875)
1863, o/c, 36.125" x 72"
Buffalo Bill Historical Center, Cody, WY
Gertrude Vanderbilt Whitney Trust Fund Purchase; 3.60

Primarily a painter of animals, Hays also succeeded in creating a splendid landscape for this work. He faced the difficult problem of conveying the immensity of the Missouri River plain. Using a high viewpoint, he made the river stretch into the far reaches of the composition.

Island Lake, Wind River Range, Wyoming
Albert Bierstadt (1830–1902)
1861, o/c, 26.5" x 40.5"
Buffalo Bill Historical Center, Cody, WY; 5.79

In 1859, Bierstadt accompanied an expedition led by Colonel Frederick Lander to the Wind River Mountains. Making oil studies and taking stereographic photographs, Bierstadt brought back to New York his first body of western imagery. *Island Lake* features the recessive planes, meticulous brushwork, and highly polished surface texture characteristic of Bierstadt's work at this time.

GOLDEN GATE, YELLOWSTONE NATIONAL PARK
THOMAS MORAN (1837–1926)
1893, O/C, 36.25" X 50.25"
BUFFALO BILL HISTORICAL CENTER, CODY, WY; 4.75

Moran's name became synonymous with Yellowstone. He accompanied the official governmental expedition into the region in 1871. His sketches of the wonders helped to convince Congress to establish Yellowstone as the first national park. The artist returned to the Park in 1892 and painted a view of the pass named Golden Gate.

Trapper Attacked by Wolves
Philip R. Goodwin (1881–1935)
c. 1905–1906, o/c, 29.25" x 15"
Buffalo Bill Historical Center, Cody, WY
Gift of Olin Corporation, Winchester Arms Collection;
26.88

In an era when the nation was "looking westward," Goodwin's sporting paintings commissioned as calendar art and advertisements for gun and ammunition companies had great appeal, especially for sportsmen and outdoorsmen fascinated by the rugged surroundings of the western wilderness. Dramatic and convincing, *Trapper Attacked by Wolves* both piques the sensual nuances and reveals the harsh, physical dangers of the frontier's winter environment.

Trapper Attacked by Wolves
Philip R. Goodwin (1881–1935)
1906, chromolithograph, 27.443" x 15.313"
Buffalo Bill Historical Center, Cody, WY
Gift of Olin Corporation, Winchester Arms Collection; 1988.8.860

Although based on Goodwin's 1905–1906 oil painting, the Winchester advertising poster made a significant change from Goodwin's earlier interpretation. While the painting depicted the trapper with a Winchester M1895 rifle, the image on the poster was altered to depict the earlier produced M1886 rifle. No explanation survives for the change.

Winchester Model 1895 Lever Action Rifle (top image)
Maker: Winchester Repeating Arms Co., New Haven, CT
Serial Number: 77,678
Caliber: .405 (.405 Winchester centerfire cartridge)
Date: 1915
Buffalo Bill Historical Center, Cody, WY
Gift of Olin Corporation, Winchester Arms Collection; 1988.8.1154

Goodwin's painting, *Trapper Attacked by Wolves*, depicts a Winchester Model 1895 similar to this rifle. Winchester began production of this rifle model in 1895. This particular rifle was produced in 1915. Page 12

Winchester Model 1886 Lever Action Rifle (bottom image)
Maker: Winchester Repeating Arms Co., New Haven, CT
Serial Number: 150,112 A
Caliber: .33 (.33 Winchester centerfire cartridge)
Date: 1910
Buffalo Bill Historical Center, Cody, WY
Gift of Olin Corporation, Winchester Arms Collection; 1988.8.3205

In the Winchester advertising poster, *Trapper Attacked by Wolves*, Goodwin depicted a Winchester Model 1886 similar to this rifle. Page 13

The Rattlesnake
Frederic Remington (1861–1909)
Copyright 1905; reworked 1908; cast c. 1910
Bronze, cast no. 17, Roman Bronze Works, NY
22.625" x 22.625" x 13"
Buffalo Bill Historical Center, Cody, WY
Gift of The Coe Foundation; 50.61

Remington used the subject of the horse reacting to the rattlesnake to create one of his most daring sculptures. He modeled the horse as a dynamic curve pulling away from the small, but deadly snake. Dissatisfied with his first version of this sculpture, Remington reworked it extensively, making this larger and more compact version.

Mule Deer in the Badlands, Dawson County, Montana
Carl Rungius (1869–1959)
1914, o/c, 59.625" x 75.25"
Buffalo Bill Historical Center, Cody, WY
Gift of Jackson Hole Preserve, Inc.; 16.93.2

For the New York Zoological Society, Rungius painted this depiction of mule deer, a buck and three does, on the edge of a canyon. In portraying the animals, his primary interest, the artist used a precise style with strong outlines. In the landscape,

Leigh painted *Panning Gold, Wyoming* late in his life, although the study for the landscape was probably painted on one of his trips to Wyoming early in the century. His figure of a miner exhibits a complicated but naturalistic pose.

View from the Mohave Wall
Wilson Hurley (b. 1924)
1976, o/c, 60.25" x 90.25"
Buffalo Bill Historical Center, Cody, WY; 6.76

Responding to the West's grandeur, Hurley follows in the tradition of the great landscape painters. He wrote, "The Grand Canyon is never the same, but like the sea, is always changing color and shadows with the hours, the daily weather, the seasons." An instantaneous vision of light on the canyon inspired this painting. Modern aids, such as photographs for accuracy of surface detail and a geological survey contour map for placement of terrain, assisted Hurley in capturing the image.

A Flight from Destiny
Bill Schenck (b. 1947)
1994, o/c, 47.375" x 62.375"
Buffalo Bill Historical Center, Cody, WY
Gift of Daniele D. Bodini; 1.95

Schenck brings characteristics of Pop Art to the western front. His hard-edged flat style with its saturated color reminds the viewer of the way that western subjects have become part of popular culture through posters, cartoons, decorative objects and other common items. He playfully recalls western myths in this painting that quotes from classic paintings by William R. Leigh and Edgar Payne.

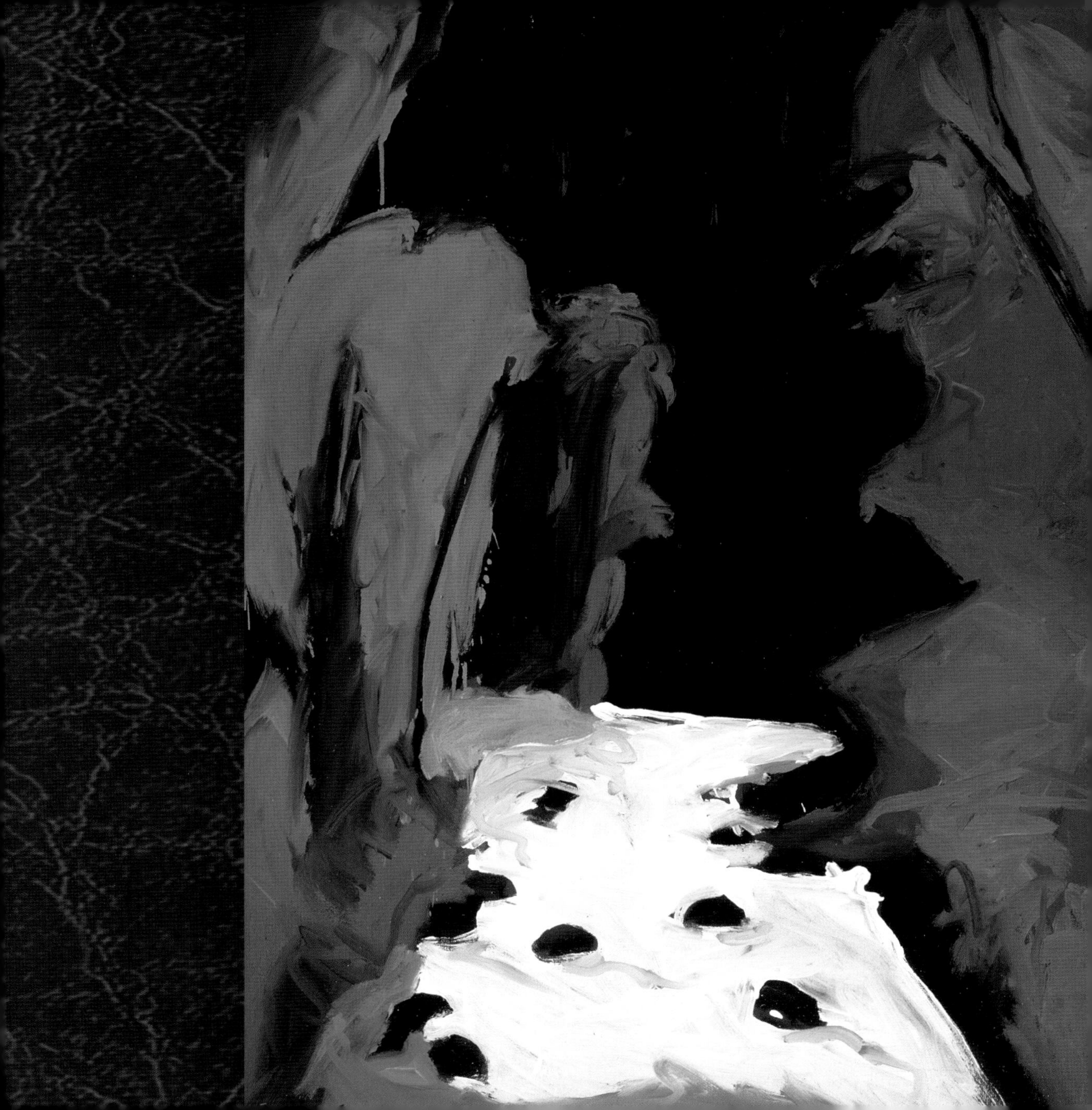

Aspen Summer (page 21)
Fritz Scholder (b. 1937)
1977, o/c, 40.25" x 30.125"
Buffalo Bill Historical Center, Cody, WY
Gift of Jack and Carol O'Grady; 25.91.2

In *Aspen Summer*, Scholder veers from the tradition of the realistically rendered landscape. Inspired by the swift, energetic vocabulary of abstract expressionism, the artist offers a contemporary, bold interpretation of the western landscape. Scholder, who is one-quarter Luiseño and was an influential instructor at the Institute of American Indian Arts, resists being regarded only as an "Indian artist" and deals with a broad range of subjects.

The Bird of Washington
John James Audubon, artist (1785–1851)
Robert Havell, Jr., engraver (1793–1878)
painted 1822, printed 1827–1838
hand colored engraving from *The Birds of America*; 37.75" x 24.75"
Buffalo Bill Historical Center, Cody, WY
Bequest of Robert D. Coe; 46.85

Audubon believed that this bird, an immature bald eagle, was a new species because of its dark plumage and called it the "Bird of Washington." The bald eagle does not acquire its white head and tail feathers until it is three or four years old.

The Bird of Washington or
GREAT AMERICAN SEA EAGLE.

Northern Hare, Winter Pelage
John James Audubon, artist (1785–1851)
R. Trembly, lithographer; Nagel & Weingaertner, printers
drawn 1841, printed c. 1851, in John James Audubon and The Rev. John Bachman,
The Quadrupeds of North America, v. 1, pl. XII (New York: V.G. Audubon, 1851–1854)
Buffalo Bill Historical Center, Cody, WY
Gift of Mrs. Robert L. Chastain

Northern Hare (old & young), Summer Pelage (page 25)
John James Audubon, artist (1785–1851)
R. Trembly, lithographer; Nagel & Weingaertner, printers
drawn 1841, printed c. 1851, in John James Audubon and The Rev. John Bachman,
The Quadrupeds of North America, v. 1, pl. XI (New York: V.G. Audubon, 1851)
Buffalo Bill Historical Center, Cody, WY
Gift of Mrs. Robert L. Chastain

John James Audubon, best known for his study of birds of America, also produced an important publication that documented the mammals of this continent. For his study of the "viviparous quadrupeds" (four-legged animals that bear their young live), he traveled to the West in 1843.

H M SHRAD

BUFFALO (PAGE 26)
HENRY MERWIN SHRADY (1871–1922)
1900, BRONZE, 12.25" X 15.5" X 6.5", CAST BY THEODORE B. STARR
BUFFALO BILL HISTORICAL CENTER, CODY, WY
GIFT OF THE COE FOUNDATION; 113.67

Shrady, a native New Yorker, studied law at Columbia University and sketched as a pastime. Encouraged to try his hand at sculpting, Shrady had his works cast by the Gorham Company and then devoted himself to the medium. He studied animals, such as the buffalo, at the Bronx Zoo. His most famous work is the Appomattox Memorial Monument to General Ulysses S. Grant in Washington, D.C.

NATURE'S CATTLE
CHARLES M. RUSSELL (1864–1926)
MODELED AND CAST 1911, BRONZE, CAST BY ROMAN BRONZE WORKS, NY
SECOND CAST OF ESTIMATED 10 LIFETIME CASTS, 4.75" X 4" X 15.325"
BUFFALO BILL HISTORICAL CENTER, CODY, WY
GIFT OF MR. AND MRS. W.D. WEISS; 27.97.7

Russell first exhibited *Nature's Cattle* in a 1911 exhibition entitled *The West That Has Passed.* "Before cows came into this country, the buffalo or bison covered the plains," wrote Nancy Cooper Russell, the artist's wife, in a commentary on this sculpture. She described it as a "family; the buffalo, the cow and the calf, trailing as though going to water."

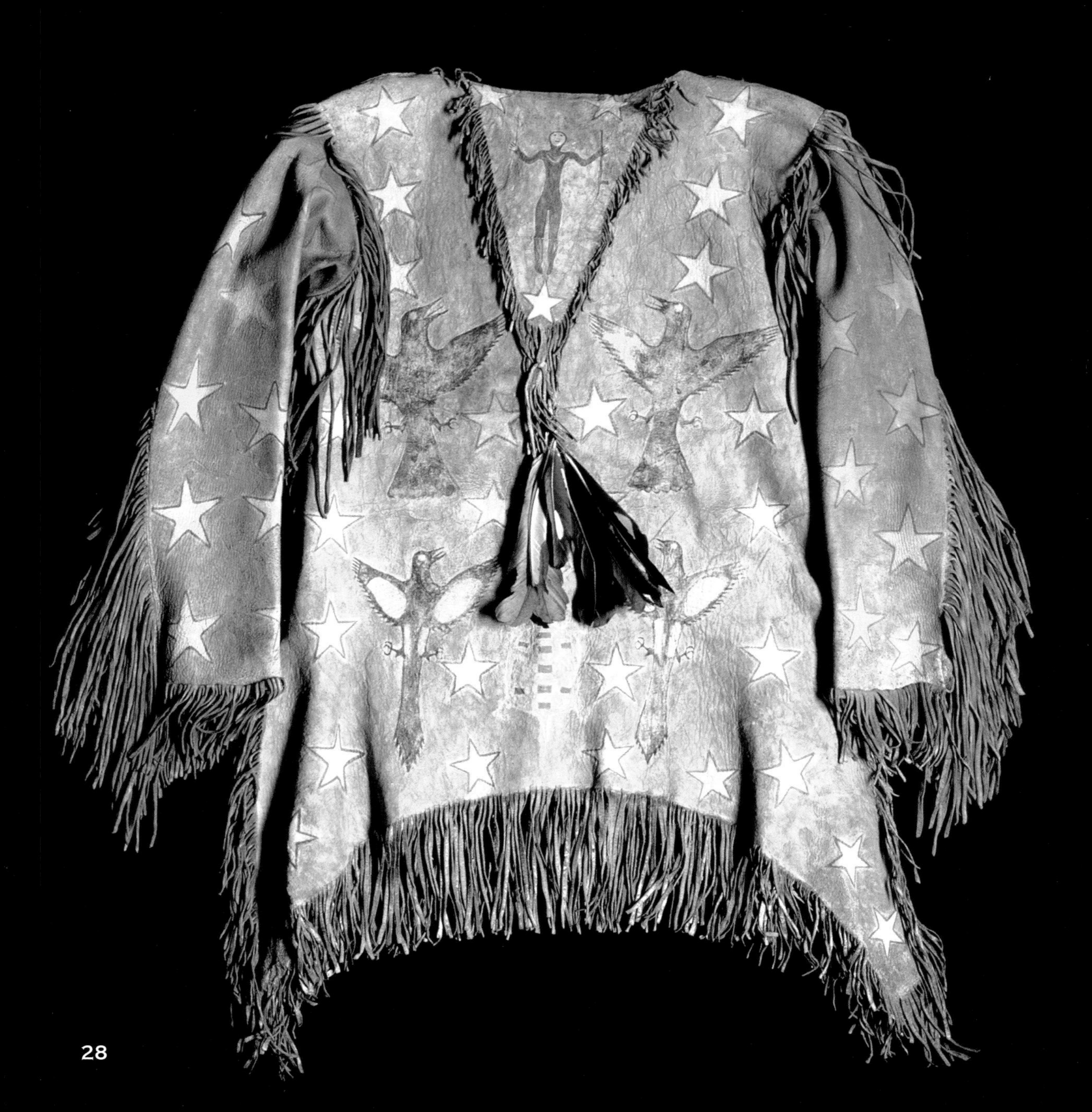

Hinono'ei (Southern Arapaho) Ghost Dance shirt, Oklahoma, c. 1890 (page 28)
elk hide, eagle feathers, pigments, 40" x 29"
Buffalo Bill Historical Center, Cody, WY
Chandler-Pohrt Collection, Gift of The Searle Family Trust and The Paul Stock Foundation; NA.204.5

Birds such as eagles, magpies, and crows serve as spiritual messengers to the heavens. The stars of the sky, the turtle, representing long life and the earth, and the bird spiritual messengers were important Ghost Dance symbols. The earth and sky together represent a whole, with elements of each including plants, animals, rocks, sun, moon, earth and people all related.

Lakota (Sioux) dress, Northern Plains, c. 1890 (page 29)
deer hide, glass beads, 63" x 58"
Buffalo Bill Historical Center, Cody, WY
Gift of Harriet D. Reed and Betty N. Landercasper, In Memory of W. Guruea Dyer; NA.202.869

Natural materials provide the body of the dress. The decorated portion, made from trade materials such as glass beads, depicts natural elements such as stars and flowers. It also provides a sense of the landscape as homeland with the abstracted tipi designs.

A Colony of Genuine Mexican Vaqueros
A. Hoen & Co., Baltimore, c. 1887
Color lithograph poster, 20.25" x 28.375"
Buffalo Bill Historical Center, Cody, WY
Gift of The Coe Foundation; 1.69.441

Posters for *Buffalo Bill's Wild West* performance often included representations of the American landscape. In this example, the bird's-eye view of the Mexican city provides the backdrop to emphasize the inclusion of Mexican vaqueros in the *Wild West. Wild West* performances featured riders representing horsemanship around the world, but the Mexican equestrians merited special attention for their influences on the United States' cowboys.

BUFFALO BILL'S WILD WEST
ND CONGRESS OF ROUGH RIDERS OF THE WORLD.

A HOEN & CO. BALTIMORE, U.S.A.

A COLONY OF GENUINE MEXICAN VAQUEROS.

ught direct at great Expense from Old Mexico by Buffalo Bill as one of the Realistic Features of his Picturesque Wild West.

A Hot Noon beside the Round-up Camp, Big Dry, Montana
L. A. Huffman (1854–1931)
1907, hand colored print
Buffalo Bill Historical Center, Cody, WY
Planned Gift of Thomas Minckler; MS-100, LAH 3626

Huffman photographed the Montana frontier from as early as 1879. His willingness to carry his large glass plate camera on horseback allowed him to capture aspects of cowboy life on the open range. This contact print was colored by Huffman.

A hot noon beside the Round-up Camp. Big Dry Montana. Neg. print coloring and copyright by

L A Huffman 07

By Nathan L. Bender, Housel Curator, McCracken Research Library

The romance of the American West is rooted in events that symbolize freedom, courage, tragedy and entrepreneurial spirit throughout the world. The early expeditions of Lewis and Clark and Zebulon Montgomery Pike brought back to the eastern United States reports of wild lands of great grassy plains; deserts, rocky mountains and rivers; of proud Indian peoples with horses; and seemingly limitless numbers of buffalo, beaver, and other wildlife. Fur traders and trappers quickly followed the early military expeditions, being the first fortune seekers from the East to experience the harsh freedoms of this western frontier. Visions of opportunity in a new land also motivated missionaries, family farmers, ranchers, miners, religious sects, outlaws and businessmen to move west, bringing their culture with them. Cattle and wagon trails were established, and the image of a pioneer family in a covered wagon crossing the Great Plains to establish a homestead of their own became firmly established as an American symbol of freedom and courage.

The Plains Indians saw the newcomers first as guests and trading partners but inevitably as a danger to their own lifestyles and customs. Conflicts led to violence, resulting in a series of battles, treaty making, treaty breaking, more battles, more treaties and more battles as the United States established its control over the lands. Peace medals were given to Native allies of the United States, continuing a tradition begun by European colonizers. The Battle of Little Big Horn is the singular western event that symbolizes the courage and tragedy of the Great Sioux War. General George Armstrong Custer, a Civil War and Plains Indian wars veteran, in June 1876, led an attack by the 7th Cavalry on an enormous tipi village of Sioux and Cheyenne peoples on the banks of the Little Big Horn River in southeastern Montana. The result was an astonishing victory by the Indians in response to a direct attack by a professional and modern army, with the result that Custer's 7th Cavalry was thoroughly decimated. The tragedy lay not only in the death of Custer and his troops, but in the fact that within two years all remaining Plains Indian peoples were confined to reservation lands and were forbidden to practice their traditional free-roaming lifestyles. Forced assimilation into American culture quickly followed, even for those nations that had been U.S. allies.

William F. "Buffalo Bill" Cody was chief of scouts for the U.S. 5th Cavalry, and an active participant in the Plains Indian wars, for which he was awarded a Medal of Honor in 1872. He was a personal friend of Gen. Custer, Wild Bill Hickok, and Chief Sitting Bull. Through association with popular writer Ned Buntline (Edward Judson), he became a celebrity of the eastern theatre and the hero of dime novel tabloids. In 1882 he organized his first *Wild West* celebration in North Platte, Nebraska, followed by a traveling *Wild West* in 1883. Eventually *Buffalo Bill's Wild West* became world famous, for Cody understood both the reality and the romance of the American West. Annie Oakley and Sitting Bull contributed to the popular success of the *Wild West*, but it was Cody's fine combination of drama and historical authenticity that recognized and nurtured the symbolic power of the American West.

Col. William F. Cody
Rosa Bonheur (1822–1899)
1889, o/c, 18.5" x 15.25"
Buffalo Bill Historical Center, Cody, WY
Given in memory of William R. Coe and Mai Rogers Coe; 8.66

Buffalo Bill enthralled Europeans with his *Wild West* exhibition when he took it to Paris in 1889. Rosa Bonheur visited the grounds of Cody's *Wild West* to sketch the exotic American animals and the Indian warriors with their families. Cody, in turn, accepted the invitation of Bonheur to visit her chateau in Fontainebleau where she painted this portrait.

INDIAN MOTHER AND CHILD
HARRY JACKSON (B.1924)
1980, BRONZE, PAINTED, CAST 3 OF EDITION OF 15, 27.75" X 35.25" X 29.75"
BUFFALO BILL HISTORICAL CENTER, CODY, WY
GIFT OF MR. AND MRS. RICHARD J. CASHMAN; 19.91

This sculpture results from Jackson's commission for a monumental sculpture of Sacagawea for a sculpture garden at the Buffalo Bill Historical Center. The historical person of Sacagawea, who traveled with Lewis and Clark, blended several roles—helpmate to the explorers, mother to her young child, victim who is restored to her family. The conflation of these roles may help to explain why Sacagawea is one of the most often portrayed women in American art.

ALEXANDER POPE.

Weapons of War (page 37)
Alexander Pope (1849–1924)
1900, o/c, 54" x 42.5" x 2"
Buffalo Bill Historical Center, Cody, WY; 201.69

From the earliest days of the frontier, artists of western subjects emphasized the authenticity and accuracy of their images. One way they demonstrated accuracy to their audience was by focusing on the replication of particular objects that signified the West. In *Weapons of War*, Pope dramatically presents an established and easily recognizable iconography of American Indian objects. Painted illusionistically, the trompe l'oeil image—artwork that literally attempts to fool the eye—engages the viewer in a confrontation of truth and art.

Custer's Last Stand
Edgar S. Paxson (1852–1919)
1899, o/c, 70.5" x 106"
Buffalo Bill Historical Center, Cody, WY; 19.69

Paxson researched the Battle of Little Big Horn and spent several years completing the painting. He then circulated it as a traveling exhibition. As part of the explanatory material on the painting, he prepared an outline key which identified the major figures.

Paxson

W. Herbert Dunton

The Custer Fight
William Herbert Dunton (1878–1936)
c. 1915, o/c, 32.25" x 50.25"
Buffalo Bill Historical Center, Cody, WY
Gertrude Vanderbilt Whitney Trust Fund Purchase; 48.61

This painting of the Battle of Little Big Horn emphasizes the fleetness and power of Plains Indian warriors. The total and decisive victory by the Sioux and Cheyenne in the face of a surprise attack by the U.S. 7th Cavalry led by a favorite Union general of the Civil War commanded the attention of the world.

Waiting for a Chinook
Charles M. Russell (1864–1926)
c. 1903, w/c on paper, 20.5" x 29"
Buffalo Bill Historical Center, Cody, WY
Gift of Charles Ulrick and Josephine Bay Foundation, Inc.: 88.60

Severe storms in the winter of 1886–1887 brought ruin to Montana's cattle industry. When owners of a herd of 5,000 cattle requested a report on their herd, cowboy Charles Russell simply drew a starving cow about to drop before ravenous wolves and titled it *Waiting for a Chinook* (a warm west wind). His drawing conveyed the impending disaster more eloquently than any written report. Later he painted this larger version and added the subtitle by which the work

This view of cattle being herded in a snowstorm was photographed by Charles J. Belden on the Pitchfork Ranch, Meeteetse, Wyoming.

PEACE MEDAL, 1849 (FRONT 8
BRONZE, BRASS BEA
BUFFALO BILL HISTORICAL CENTER, CO
GIFT OF GOELET GALLATIN; NA.2

peace medal features U.S. President Zachary Taylor. Medals such as this were given to high-ranking leaders o
rican Indian nations as symbols of peace and friendship. They served to recognize and mark the important
ion of their recipients, making them highly valued possessions.

SHAR-I-T
HENRY INMAN (1801–
1832, O/C, 30.375" X 25
BUFFALO BILL HISTORICAL CENTER, CO
GIFT OF THE BLANK FAMILY FOUNDATION; 2

I-Tar-Ish, a Pawnee chief, visited Washington, D.C. in 1821 and had his portrait painted by Charles Bird King. No
medal about his neck, symbol of his friendship with the United States of Americ. Later, Henry Inman painted cop
s Indian portraits so that lithographers could make prints based on the paintings. Most of King's Indian portraits h

The Summit Springs Rescue, 1869 (detail)
Charles Schreyvogel (1861–1912)
1908, o/c, 48" x 66"
Buffalo Bill Historical Center, Cody, WY
Bequest in memory of Houx and Newell families; 11.64

William F. Cody was a major participant in the Battle of Summit Springs, an event that he dramatically re-enacted in *Buffalo Bill's Wild West*. Schreyvogel's painting of the actual rescue by the U.S. 5th Cavalry of kidnapped women held in a Cheyenne village in 1869 is based on his interviews with Cody and viewings of the *Wild West* re-enactments.

W H D
Koerner

Madonna of the Prairie (page 48)
W.H.D. Koerner (1878–1938)
1921, o/c, 37" x 28.75"
Buffalo Bill Historical Center, Cody, WY; 25.77

In the novel, *The Covered Wagon*, Molly Wingate traveled the Oregon Trail with a wagon train of settlers. Encountering prairie fires and Indian arrows, the beautiful maiden eventually reached Oregon, where, in the conventions of popular fiction, she found true love. In this illustration for the novel's dust jacket, Koerner used the covered wagon to form a halo around the pioneer's head.

The Prairie Burial (page 49)
William Tylee Ranney (1813–1857)
1848, o/c, 28.5" x 41"
Buffalo Bill Historical Center, Cody, WY
Gift of Mrs. J. Maxwell Moran; 3.97

The Prairie Burial chronicles a tragic aspect of western migration. Ranney's representation of a grieving family standing before a small grave serves as a reminder of the many children who died due to illness and accidents during the period of western expansion.

Oglala Lakota (Sioux) shirt, Northern Plains, c. 1885
Tanned deer hide, glass beads, human hair, ermine, wool cloth,
feathers, porcupine quills with pigment; 37"
Buffalo Bill Historical Center, Cody, WY
Gift of Mr. and Mrs. Robert Maxwell James; NA.202.208

Officers of tribes, sometimes called "shirt wearers," were presented with decorated hide shirts to wear during their tenure. It was a privilege to wear such a garment, and such privilege had to be earned by the shirt owners.

ÜBER

DIE SELBSTSTÄNDIGKEIT DER SPECIES

DES

URSUS FEROX Desm.

VON

PRINZ MAX VON WIED,

MIT

ANATOMISCHEN BEMERKUNGEN

VON

Dr. C. MAYER,

MM. d. A. d. N.

MIT DREI STEINDRUCKTAFELN.

DER AKADEMIE ÜBERGEBEN DEN 31. MÄRZ 1856.

LIFE ON THE PLAINS,
BY GEN. G. A. CUSTER
U·S·ARMY

Über die Selbstständigkeit der Species des Ursus Ferox, Desm. 1856 (page 52)
Prinz Max von Weid (1782–1867) and Dr. C. Mayer
[Bonn]: Mit drei Steindrucktafeln. Der Akademie ubergeben den 31. Marz 1856. v. XXVI. P.I.
Buffalo Bill Historical Center, Cody, WY

This is one of the earliest scientific descriptions of the grizzly bear. Prince Alexander Philip Maximilian was son of the Prussian ruler Friedrich Karl of Wied-Neuwied and a trained naturalist in his own right. He traveled the upper Missouri River in 1833–34 and collected much information along the way, thirty years after the Lewis and Clark Expedition.

Life on the Plains, or Personal Experiences with Indians, 1874 (page 53)
Gen. George Armstrong Custer (1839–1876)
New York: Sheldon and Company, 1874.
Buffalo Bill Historical Center, Cody, WY

This autobiographical account of General Custer described his martial exploits during the western Plains Indian wars. It was published just two years before his death at the Battle of Little Big Horn.

Holograph letter to Gen. George Crook, 12 pages on 3 sheets folded
Gen. Alfred H. Terry (1827–1890)
Camp on north side of Yellowstone River near mouth of Big Horn, July 9, 1876
Buffalo Bill Historical Center, Cody, WY; MS-7

This letter from Gen. Terry, commander of the Department of Dakota, to Gen. Crook of the U.S. 5th Cavalry is a field report explaining the circumstances of Gen. Custer's defeat at Little Big Horn. The following quote is from page two:

"I greatly regret to say that Custer and every officer and man under his immediate command were killed. Reno was driven back to the bluffs, where he was saved by the remainder of the regiment. He was surrounded by the enemy and was obliged to entrench himself, but succeeded in maintaining himself in his position, with heavy loss until the appearance of General Gibbon's column induced the Indians on the evening of the 26th ultimo to withdraw. Two hundred and sixty eight officers, men and civilians were killed and there were fifty two wounded."

of five Companies, the other under Major Reno of three Companies. The attacks of these two detachments were made at points nearly or quite three miles apart.

I greatly regret to say that Custer and every Officer and man under his immediate Command were Killed. Reno was driven back to the bluffs, where he was joined by the remainder of the regiment.

He was surrounded by the Enemy and was obliged to entrench himself, but succeeded in maintaining himself in this position with heavy loss until the appearance of General Gibbons Column induced the Indians, on the Evening of the 26th ultimo to withdraw.

Two hundred and fifty Eight Officers men and Civilians were Killed and there were fifty two wounded.

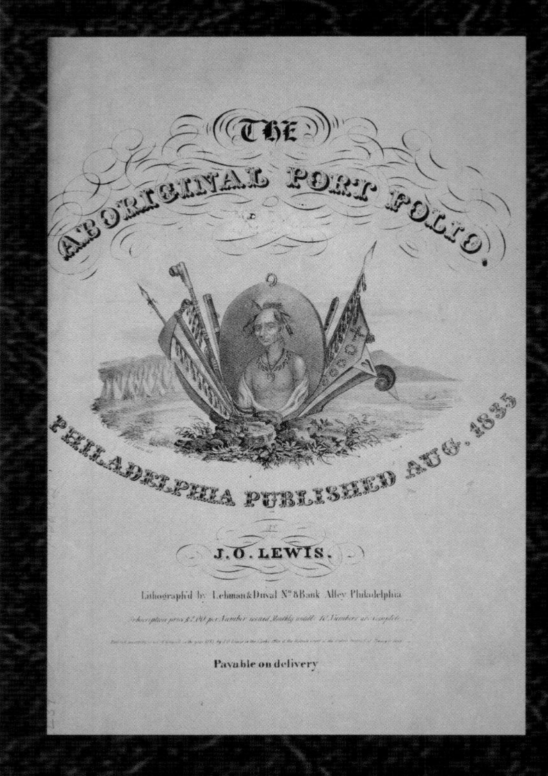

Waa-Pa-Laa or The Playing Fox, in *The Aboriginal Portfolio*
James Otto Lewis (1799–1858)
[Plate 8] taken at Prairie du Chien by J.O. Lewis 1825
Philadelphia: Lithographed by Lehman & Duval, 1835
Buffalo Bill Historical Center, Cody, WY
Gift of Clara S. Peck; 26.71

A Sac chief, Waa-Pa-Laa, played prominent roles in the War of 1812 and the later Black Hawk War of 1832 against the United States. In 1825 he participated in making the Treaty of Prairie du Chien that established tribal boundaries for the Sac and Fox, Ojibwa, Santee Sioux, Winnebago, Potawatomi and Iowa of the upper Mississippi River region.

HISTORY

OF

THE INDIAN TRIBES

OF

NORTH AMERICA,

WITH

BIOGRAPHICAL SKETCHES AND ANECDOTES

OF THE

PRINCIPAL CHIEFS.

EMBELLISHED WITH ONE HUNDRED AND TWENTY PORTRAITS,

FROM THE

INDIAN GALLERY

IN THE

DEPARTMENT OF WAR, AT WASHINGTON.

BY THOMAS L. M'KENNEY,
LATE OF THE INDIAN DEPARTMENT, WASHINGTON,
AND JAMES HALL, ESQ.
OF CINCINNATI.

VOLUME I.

PHILADELPHIA:
PUBLISHED BY EDWARD C. BIDDLE,
23 MINOR STREET.
1837.

Mo-Hon-Go an Osage Woman, in *History of the Indian Tribes of North America, with Biographical Sketches and Anecdotes of the Principal Chiefs*
Thomas L. McKenney (1785–1859) and James Hall (1793–1868)
V.1, p. 44
Philadelphia: Edward C. Biddle, 1837
Buffalo Bill Historical Center, Cody, WY
Gift of Clara S. Peck; 42.70.4

Mo-Hon-Go was part of a small group of Osage Indians tricked into traveling to Europe in 1827 where they were exhibited, eventually catching the attention of French high society. Lafayette arranged for their return to Washington, D.C., where Mo-Hon-Go had her portrait painted by Charles Bird King. This lithograph was drawn, printed and colored by J.T. Bowen of Philadelphia for the publication of the McKenney and Hall album.

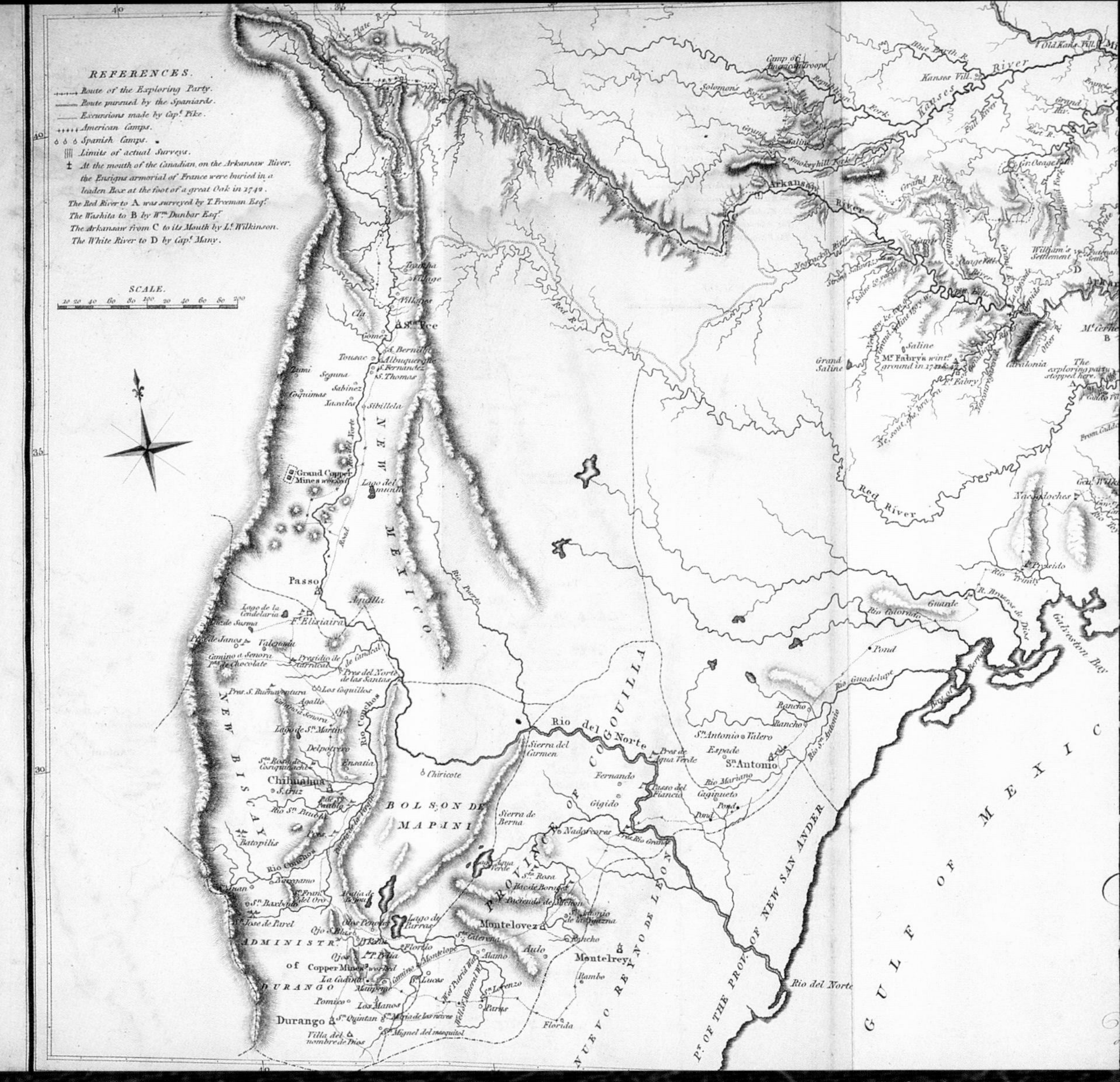

Captain Zebulon Montgomery Pike led the first U.S. military exploratory expedition of the Mississippi River headwaters in 1805. In 1806–07 he led another expedition to the discovery of Pike's Peak in Colorado while exploring the headwaters of the Arkansas River and the Spanish southwest. His travels resulted in the first American maps of the southwestern territories.

EXPLORATORY TRAVELS

THROUGH THE

WESTERN TERRITORIES

OF

NORTH AMERICA:

COMPRISING A

VOYAGE FROM ST. LOUIS, ON THE MISSISSIPPI,

TO THE

SOURCE OF THAT RIVER,

AND A

JOURNEY THROUGH THE INTERIOR OF LOUISIANA,

AND THE

NORTH-EASTERN PROVINCES OF NEW SPAIN.

Performed in the years 1805, 1806, 1807, by Order of the Government of the United States.

BY ZEBULON MONTGOMERY PIKE,

MAJOR 6TH REGT. UNITED STATES INFANTRY.

LONDON:

PRINTED FOR LONGMAN, HURST, REES, ORME, AND BROWN,
PATERNOSTER-ROW.

1811.

EXPLORATORY TRAVELS THROUGH THE WESTERN TERRITORIES OF NORTH AMERICA...
CAPT. ZEBULON MONTGOMERY PIKE (1779–1813)
FRONTISPIECE "MAP OF THE INTERIOR OF LOUISIANA, WITH A PART OF NEW MEXICO."
LONDON: LONGMAN, HURST, REES, ORME, AND BROWN, 1811.
BUFFALO BILL HISTORICAL CENTER, CODY, WY

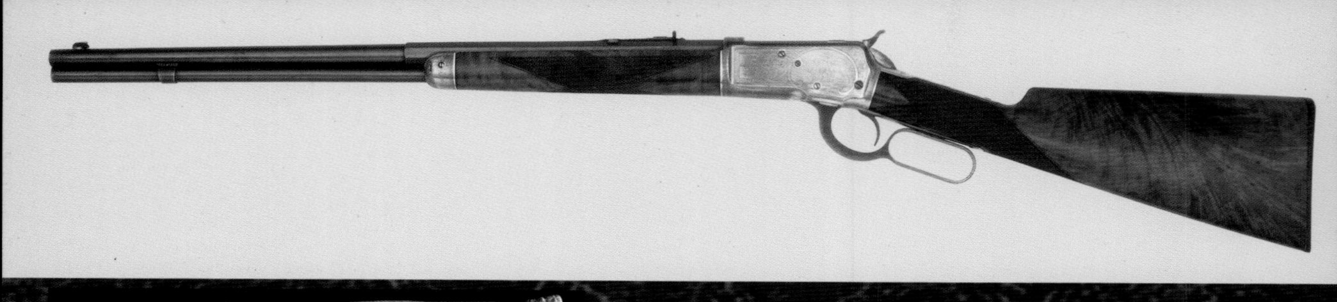

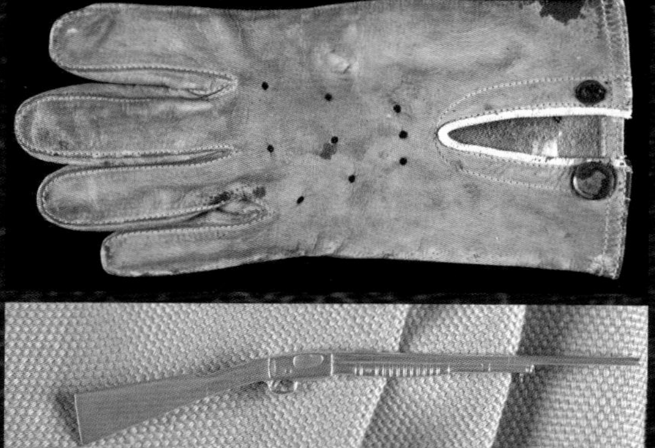

Annie Oakley's Model 92 carbine in .32 WCF serial #41023
Winchester factory engraved with gold finish, by J. Ulrich
Winchester Repeating Arms Co., New Haven, CT, 1895
Buffalo Bill Historical Center, Cody, WY
Gift of Mr. and Mrs. Spencer T. Olin (Anne); 1.69.1866

This rifle was a custom order for *Buffalo Bill's Wild West* sharpshooter Annie Oakley (1860–1926). It features her initials engraved on the left side of the receiver and her name stamped on both sides.

Annie Oakley's glove
Leather with brass snap
Buffalo Bill Historical Center, Cody, WY
Gift of Adele Von Ohl Parker; 1.69.66

Used by Annie Oakley in her sharpshooting exhibitions of *Buffalo Bill's Wild West*, this glove has the name ANNIE OAKLEY hand printed on its upper palm side. Adele Von Ohl Parker, the donor, knew Annie as a fellow performer in the *Wild West.*

Remington rifle pin
10K gold, 1918
Buffalo Bill Historical Center, Cody, WY
Gift of Mr. and Mrs. Alan Lovelace (Melanie); 1.69.2337

This pin belonged to Annie Oakley, famous woman sharpshooter of *Buffalo Bill's Wild West.* It was given to Annie on March 23, 1918 at Pinehurst, North Carolina, where she often gave shooting lessons.

Miss Annie Oakley, The Peerless Lady Wing-shot
A. Hoen & Co.
Color lithograph poster
Baltimore, Maryland, c. 1890
Buffalo Bill Historical Center, Cody, WY
Gift of the Coe Foundation; 1.69.73

Annie Oakley (1860–1926) began shooting with *Buffalo Bill's Wild West* in 1885 with her husband Frank Butler and became world famous as a result.

BUFFALO BILL'S WILD WEST·

CONGRESS, ROUGH RIDERS OF THE WORLD.

MISS ANNIE OAKLEY,

THE PEERLESS LADY WING-SHOT.

Annie Oakley signed cabinet card
Stacy
Brooklyn, New York, n.d.
Buffalo Bill Historical Center, Cody, WY;
Vincent Mercaldo Collection; P.71.362.1

This is an early studio photograph produced as a cabinet card, showing Annie Oakley holding a double-barrelled shotgun over her shoulder and standing next to a case full of shooting medals.

SITTING BULL CABINET CARD
WILLIAM NOTMAN AND SON
MONTREAL, CANADA, 1885
BUFFALO BILL HISTORICAL CENTER, CODY, WY: P.69.1578

Sitting Bull (1834–1890) was the Hunkpapa Sioux medicine man and chief who in the late 19th century was commonly credited for the victory at Little Big Horn. In 1885 he was asked by William F. Cody to appear in his *Wild West*. There he joined Annie Oakley, also in her first year with the show. Sitting Bull called her “Little Sure Shot,” while she came to look upon him as her “adopted father,” as is written on the front of this cabinet card. On the verso of this card is written “return to Annie Oakley.”

Compliments Of Ann

iokler

Bridle and Martingale
1887, Black leather with cowrie shells and brass hardware
Buffalo Bill Historical Center, Cody, WY; 1.69.43 A/B

A gift to William F. Cody from the Prince of Wales in 1887, this bridle and martingale horse tack quickly became favorites and were used extensively in *Buffalo Bill's Wild West* and its associated publicity.

"The Farewell Shot" Positively the last appearance of Col. W.F. Cody (in the saddle), "Buffalo Bill."
U.S. Lithograph Co.
Color lithograph poster, Russell-Morgan, Print; 28" x 41"
Cincinnati and New York, c. 1910
Buffalo Bill Historical Center, Cody, WY; 1.69.137

This colorful poster of *Buffalo Bill's Wild West* features William F. Cody (1846–1917) in one of his beaded buckskin jackets on a white horse bedecked with the black leather and cowrie shell bridle and martingale given to him by the Prince of Wales.

"THE FAREWELL SHOT"
POSITIVELY THE LAST APPEARANCE
OF
COL. W. F. CODY, (IN THE SADDLE)
"BUFFALO BILL"

Buffalo Bill's beaded buckskin jacket
c. 1898, Tanned buckskin, glass beads, brass buttons, blue satin
Buffalo Bill Historical Center, Cody, WY
Gift of Mr. Robert Garland; 1.69.784

One of William F. Cody's buckskin jackets, this heavily fringed jacket features the American and Cuban flags beaded onto its back. Adopting a style from American Indians, beaded buckskin clothing was known and used by American frontiersmen.

Buffalo Bill's "No. 1" Military Rifle in .43 Spanish, serial #3
E. Remington & Sons
Factory engraved, gold plated hammer and block, checkered walnut stock.
Ilion, New York, 1872–73
Buffalo Bill Historical Center, Cody, WY
Gift of Mr. and Mrs. Harry Schloss in memory of his grandfather, Moses Kerngood; 1.69.2412

Cody was presented with this Remington-Rider rolling block rifle as a gift from Elisha H. Greene during his visit to the Remington factory. He used the rifle during his theatrical stage play *Scouts of the Plains* with his friends "Wild Bill" Hickok and "Texas Jack" Omohundro. Cody later gave this rifle to his good friend Moses Kerngood, c. 1875.

CUSTER'S LAST STAND

SENTED BY
FFALO BILL'S
WILD WEST.

Hawken .56 caliber half stock percussion rifle
Jacob (1786–1849) and Samuel (1792–1884) Hawken
St. Louis, Missouri, c. 1825–45
Buffalo Bill Historical Center, Cody, WY
Gift of William B. Ruger, Sr. and Sturm, Ruger & Company; 1997.4.2

The muzzleloading Hawken rifles were among the finest known to western mountain men during the late fur trade era. They were famous for their accuracy and reliability. The St. Louis gunshop of brothers Jacob and Samuel Hawken employed over a dozen men and could produce about 100 rifles a year. Hawken rifles were used by many famous mountain men including Jim Bridger, Kit Carson and John "Liver Eating" Johnston. This rifle features a 331/4" barrel with seven groove rifling and is signed "J. and S. Hawken." The lock is marked "Kingsland & Co., Warranted."

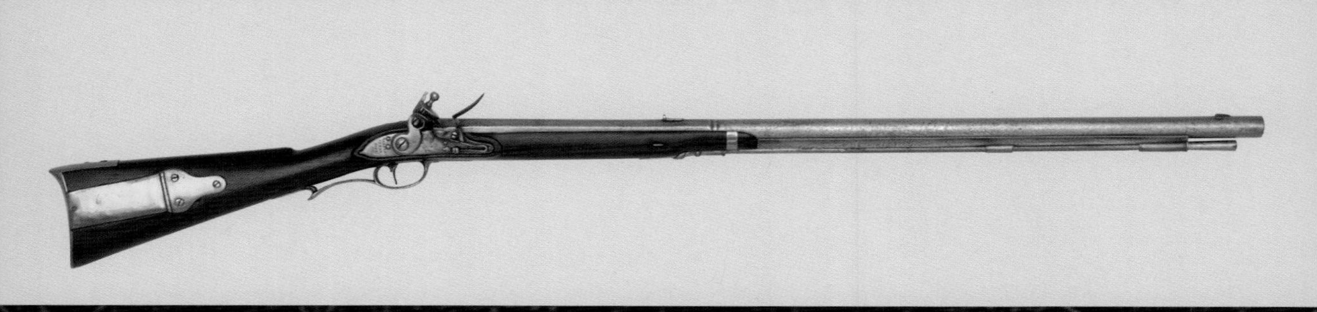

U.S. M1803 Flintlock Rifle, serial #94
Harpers Ferry Armory
Harpers Ferry, Virginia, 1804
Buffalo Bill Historical Center, Cody, WY
Gift of Olin Corporation, Winchester Arms Collection; 1988.8.1584

The M1803 was the first military rifle manufactured at an armory of the United States government. Lewis and Clark were possibly equipped with prototypes of this model, but production of the standard form shown here did not commence until the spring of 1804 due to difficulties in barrel manufacture, too late for use by Lewis and Clark who started in 1804. However, it is quite possible that the standard M1803 rifles were used by the Pike expedition of 1806. This rifle has a .54 caliber 33" barrel, the length of which was increased to 36" on rifles manufactured after 1807.

Lukens (Butt Reservoir) Air Rifle
Isaiah Lukens; c. 1800–1810
Philadelphia, PA
Buffalo Bill Historical Center, Cody, WY; 1991.5.1

This is one of two known Lukens rifles with "LUKENS" marked on the lockplate. One of these "large" Lukens air guns was made for and used by the Lewis and Clark expedition, and it is mentioned in their journals at least 16 times. The rifle was powered by pumping air into its hollow shoulder stock. Once sufficient pressure was obtained, multiple discharges could be fired with considerable force.

U.S. M1873 Carbine, serial #34962
Springfield Armory
Springfield, Massachusetts, c.1876
Buffalo Bill Historical Center, Cody, WY; 1.69.2178

This "trapdoor" single-shot .45 caliber carbine model was standard issue to U.S. Cavalry units during the Indian wars after 1873. This carbine, serial number 34962, may have been issued to the U.S. 7th Cavalry. Ammunition was the

GEORGE L SCHWARZ

Buffalo Bill's Wild West Show Personnel, c. 1912
Black and White Photograph, 11" x 64.375"
Buffalo Bill Historical Center, Cody, WY; P.69.16

The People of the American West

By Warren Newman, Cody Firearms Museum Curatorial Assistant

Mental images of the people of the early American West are dramatically vivid and virtually universal. They have been implanted and nurtured in the minds of children and adults alike, not only in this country but around the world, by sensational paperback novels and action-packed "western" movies, by magazine illustrators and commercial artists, by flamboyant entertainers and imaginative historians. These cognitive images are remarkably similar. They usually depict hearty families braving the rigors and uncertainties of a grand but untamed land in creaking covered wagons; hostile Indians intent upon the massacre of the would-be settlers and bands of military cavalry galloping to the rescue; bearded gold prospectors leading pack mules up a mountain trail; the thundering hooves and daring riders of a homestead land rush; and, perhaps most pervasively, hard-living, horse-riding, cattle-herding, quick-shooting cowboys.

These familiar images, while constituting both a popular myth and a national symbol, are a fractional and inadvertently deceptive part of the story of the people of the West. The greater reality is much more complex. The people of the West are, in fact, not readily stereotyped. Native Americans were nations whose cultural identities remained distinctive and relatively intact regardless of interaction with others and enforced assimilation pressures. The same is true of the vast array of immigrants whose heritages can be clearly traced to their countries of origin in Europe and Asia and Latin Americ. The greatly disparate nature of occupational pursuits and social and educational backgrounds broadens an over-simplified perception of Westerners. They were farmers, hunters, trappers, shopkeepers, ranchers, doctors, lawyers and a profusion of other callings, not to mention gamblers, gunmen, cattle rustlers and horse thieves. The motivations for their westward sojourns were diverse. Some were running away from trouble, failure or frustration. Others were in search of wealth, happiness or adventure. Still others were stirred by the anticipation and challenges of a new life in a new land. In short, they were as different and as complicated and as unique as the people who remained behind. To relegate them to simplistic categories is to distort and confuse our perception of them.

There are, however, some generalizations that are helpful in understanding the people of the American West. Prominent among them is that, for the most part, Westerners were people of perseverance and tenacity. They found the will to keep going long after bodies were exhausted, minds had lost their clarity, and any hope of survival or success had faded into obscurity. Zane Grey's repeated early failures as a writer only spurred him to travel westward to gain new ideas and develop innovative techniques until he began to attract a following of readers and, ultimately, become one of the most successful and prolific authors of his era.

Another generalization that aids the understanding of the people of the West is that they were, predominantly, unfailing optimists. They were sustained incessantly by the continuing expectation that better things, greater beauty, and more satisfying lives lay ahead of them. They were constantly charmed by the lure of both near and distant horizons.

Few generalities about them are more legitimate and apparent than that Westerners were people of astonishing courage. Not the kind of foolhardy bravery in which there is no fear, but the deeper kind of physical and spiritual courage that experiences fear in all its dreadful dimensions, yet enables people to forge ahead and do what has to be done to get the job done, or win the battle, or finish the journey. William F. "Buffalo Bill" Cody epitomizes that spirit. Cody was an intrepid Pony Express rider, a remarkable hunter, and an Army scout who served with such distinction that he was awarded the Congressional Medal of Honor in 1872. He embodied and reflected the spirit of the words that had been used in 1813 by President Thomas Jefferson to describe Meriwether Lewis when he wrote that Lewis was "Of undaunted courage…"

These distinguishing features were among a number that the people of the early American West molded into the character of a fledgling nation. To paraphrase the words of writer Wallace Stegner, "The West *is* America, only more so."

The Broncho Buster
Frederic Remington (1861–1909)
1895, bronze, cast no. 21, 23.375"
cast in 1895 by Roman Bronze Works
Buffalo Bill Historical Center, Cody, WY; Gift of
G.J. Guthrie Nicholson Jr. and son in memory of their father/grandfather
G.J. Guthrie Nicholson, rancher at Four Bear, Meeteetse, WY; 7.74

This sculpture was Remington's first experiment in bronze. So vital and energetic was the piece and so untraditional in its approach, it won for him immediate recognition as a sculptor and helped him enter the formal ranks of American artists of the day. Remington was enthusiastic about the potential for sculpture and wrote to Owen Wister in 1895, "...all other forms of art are trivialities—mud—or its sequence 'bronze' is a thing to think of when you are doing it—and afterwards too. It dont [sic] decay. The moth dont [sic] break

The Conquest of the Prairie
Irving R. Bacon (1875–1962)
1908, o/c, 47.25" x 118.5"
Buffalo Bill Historical Center, Cody, WY
Bequest in memory of Houx and Newell families; 14.64

This painting portrays William F. "Buffalo Bill" Cody as the guide bringing modern life to the West. In the distance gleams the future—an industrialized city. Cody purchased this painting from the artist and displayed it in the Irma Hotel where it hung until it was donated to the Buffalo Bill Historical Center.

Self Portrait
Charles M. Russell (1864–1926)
1900, w/c on paper, 12.375" x 6.875"
Buffalo Bill Historical Center, Cody, WY
Gift of Charles Ulrick and Josephine Bay Foundation, Inc.; 98.60

With his feet planted solidly and his hat tipped back, Russell portrays himself as a stalwart, yet open, person. He wears the red Metis sash that marked his individuality. As he wrote about himself, "I am old-fashioned and peculiar in my dress. I am eccentric (that is a polite way of saying you're crazy). I believe in luck and have had lots of it.... Any man that can make a living doing what he likes is lucky, and I'm that."

SITTING BULL KILLING A CROW INDIAN
SITTING BULL (1831–1890)
1882, PENCIL AND CRAYON ON PAPER, 8.5" X 10.25"
BUFFALO BILL HISTORICAL CENTER, CODY, WY: 40.70.2

Sitting Bull (Tatanka Iyotanka), a Hunkpapa Lakota and perhaps the most famous of all American Indians, received early recognition from his tribe as a warrior and man of vision. In addition, he was an artist, recording historical accounts and depicting feats in battle through ledger drawings—visual works drawn on blank sheets of ledger books obtained from U.S. soldiers, traders, missionaries and reservation employees.

N.C.WYETH

The Scout (page 82)
Harvey Dunn (1884–1952)
1910, o/c, 36.125" x 24.125"
Buffalo Bill Historical Center, Cody, WY; 5.77

Born and raised in a sod house on the South Dakota prairie, Dunn drew upon his surroundings for inspiration in painting scenes of prairie life and toil in the early days of settlement. In the early 1900s, he became a student of Howard Pyle, one of America's foremost illustrators at that time. Many of Dunn's subjects were painted for books and magazine illustrations. *The Scout* appeared on the cover of *Outing* magazine in 1910.

The Broken Bow (page 83)
Joseph Henry Sharp (1859–1953)
c. 1912, o/c, 44.5" x 59.375"
Buffalo Bill Historical Center, Cody, WY; 7.75

In Montana, Sharp lived in a log cabin; in New Mexico, he lived in an adobe house and used an abandoned chapel for his studio. He probably painted this work in his Taos studio but posed his models wearing Plains Indian clothing.

Cutting Out
N.C. Wyeth (1882–1945)
1904–1905, o/c, 38" x 25.875"
Buffalo Bill Historical Center, Cody, WY
Gift of John M. Schiff; 45.83

Wyeth wrote, "Cutting out is a hard, wearisome task. There were some six thousand cattle in the herd that had been rounded up that morning, and it was the work of the men to weave through that mess and drive out certain brands . . . ."

A Contemporary Sioux Indian
James Bama (b. 1926)
1978, oil on panel, 23.375" x 35.375"
Buffalo Bill Historical Center, Cody, WY
William E. Weiss Contemporary Art Fund Purchase; 19.78

In western art, the Indian is often portrayed as a character from the past. Using a realist style, Bama portrayed a contemporary Indian who maintains a relationship with the past but has to find his place in the white man's world. The message on the wall behind the subject echoes the artist's theme of the nonacceptance of Indians in mainstream American society.

The Buffalo Hunt
Frederic Remington (1861–1909)
1890, o/c, 34" x 49"
Buffalo Bill Historical Center, Cody, WY
Gift of William E. Weiss; 23.62

In the fall of 1890, Remington made his yearly sojourn to the West, this time to Montana and the Big Horn Mountains, to gather material and inspiration for his art. The graphic realism and action of *The Buffalo Hunt* exemplifies his early style. Remington turned a typical event in the life of the Indian into a moment of tragedy and drama, reminding the viewer of the constant dangers of life in the West.

Appeal to the Great Spiri
Cyrus E. Dallin (1861–194
1913, bronze, 21" x 21.5" x 15", cast no. 92, Gorham Co. Founder
Buffalo Bill Historical Center, Cody, W
Gift of Laurance S. Rockefeller; 10.0

Appeal to the Great Spirit represented, in Cyrus Dallin's view, "glorification of the red man's 'lost cause.' Resistance havin
proved as vain as the overtures of peace and the vaticinations of the prophet, there is nothing left for him but 'an appeal t
he higher court.'" With its smoothly rounded form and surface, and the representation of the Indian with outstretched arm

The Bucker and the Buckeroo
Charles M. Russell (1864–1926)
modeled 1923–1924, copyright 1925
bronze, 14.75" x 8" x 10.5"
cast by Roman Bronze Works, NY
Buffalo Bill Historical Center, Cody, WY
Gift of William E. Weiss; 9.81

In this sculpture, Russell works most dynamically with the full implications of three dimensions. Many of his sculptures are oriented pictorially and, although modeled in the round, exist essentially in one plane. The horse and rider in this sculpture twist and turn, occupying space fully and giving the viewer multiple vantage points. *The Bucker and the Buckeroo* was Russell's title for this work, although it has sometimes been called *The Weaver.*

DOUGLAS DUER

White Man Runs Him-Apsaroke (page 92)
Edward S. Curtis (1868–1952)
Photogravure from Curtis copyright photograph 1908
Portfolio volume 4, plate 115, from *The North American Indian*, set #329/500, Cambridge, Ma: University Press, 1907–1930
Buffalo Bill Historical Center, Cody, WY
Gift of Douglas L. Manship, Sr.

End of the Trail (page 93)
James Earle Fraser (1876–1953)
modeled 1915, cast 1918,
bronze, 33.75" x 26" x 8"
cast no. 12, Roman Bronze Works, NY
Buffalo Bill Historical Center, Cody, WY
Clara Peck Purchase Fund; 112.67

End of the Trail has appealed to public sentiment since its conception following Chicago's 1893 World's Columbian Exposition. Inspired by the exposition's fusing of the nostalgic with the progressive, Fraser made a windblown and destitute symbol that represented the public's belief in the sad, yet inevitable extinction of the Indian. In 1915 the artist exhibited a monumental version of this subject at the thematically-appropriate site of the Panama-Pacific International Exposition in San Francisco, and its popularity led him to make smaller casts such as this one.

Wildfire, book
Zane Grey
The Mysterious Rider, book
Zane Grey
Buffalo Bill Historical Center, Cody, WY
Frank Tenney Johnson Collection; MS-12

Zane Grey's Winchester M1895 Sporting Rifle
(presented to Zane Grey by the Winchester Repeating Arms Company on February 28, 1924)
Maker: Winchester Repeating Arms Co., New Haven, CT
Engraver: Philip Clunt
Caliber: .30-06
Buffalo Bill Historical Center, Cody, WY
Donated in loving memory of Robert Jesse Moore by his family; 1991.1.1

Born 1872 in Zanesville, Ohio, Grey had a great interest in baseball, receiving a scholarship in the sport from the University of Pennsylvania. He graduated in 1896 in the field of dentistry, but he continued to play amateur baseball for several years. In 1900 he met his future wife, Line Elise Roth, who encouraged him in his writing, but he had little success with it initially. He traveled to the West, gaining new ideas for his books, and gradually developing a following of readers. By 1915 he had published fifteen books and was widely recognized for his work. He moved his family to California in 1918, which allowed him to begin to work closely with the newly-emerging film industry. There he became a prolific writer, publishing an average of one book a year for most of the remainder of his career.

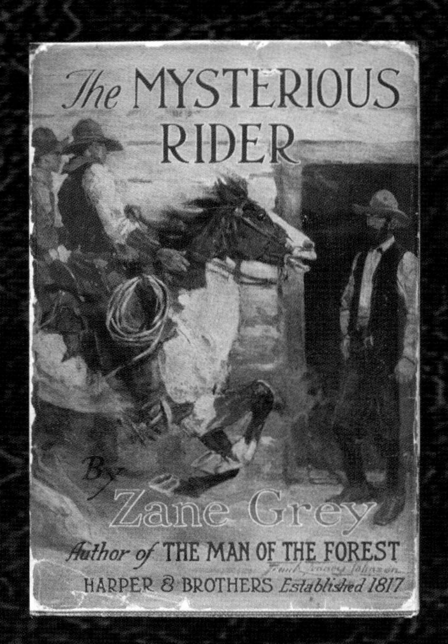
The MYSTERIOUS RIDER
By
Zane Grey
Author of THE MAN OF THE FOREST
HARPER & BROTHERS Established 1817

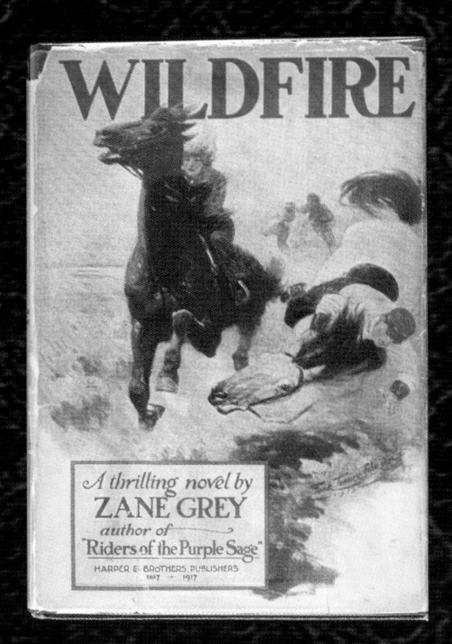
WILDFIRE
A thrilling novel by
ZANE GREY
author of
"Riders of the Purple Sage"
HARPER & BROTHERS, PUBLISHERS
1817 1917

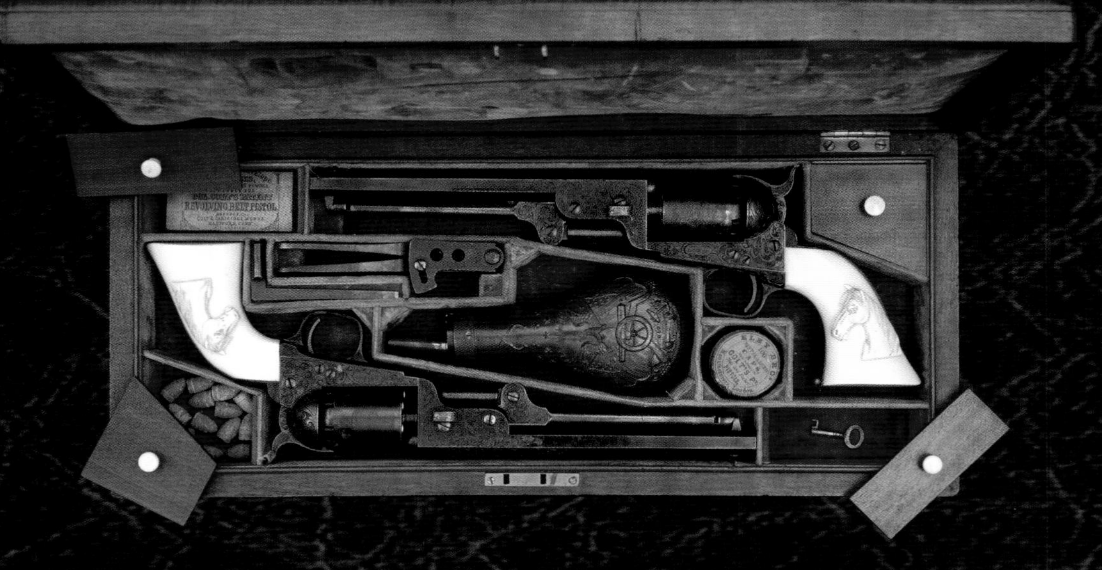

Colt Model 1851 Percussion Revolvers
Maker: Colt's Patent Firearms Mfg. Co., Hartford, CT
Serial Number: 96,456 & 96,458
Caliber: .36
Date: 1860
Buffalo Bill Historical Center, Cody, WY
Gift of James R. Woods Foundation; 1979.4.1.2

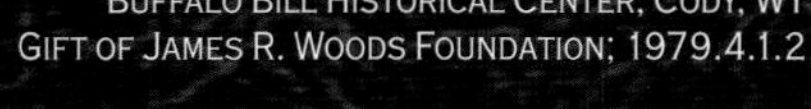

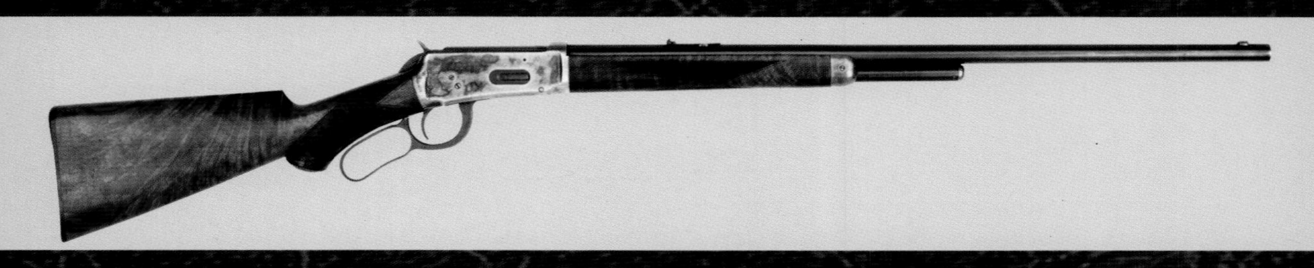

Frederic Remington's Winchester M1894 Deluxe Sporting Rifle
Maker: Winchester Repeating Arms Co., New Haven, CT
Serial Number: 17,672

Engraved Colt Model 1849 "Pocket" 4.5" Barrel Percussion Revolver,
1st Type (Cased)
Maker: Colt's Patent Firearms Mfg. Co., Hartford, CT
Serial Number: 110,281
Caliber: .31 (7 groove rifling)
Date: 1856
Buffalo Bill Historical Center, Cody, WY
gift of Olin Corporation, Winchester Arms Collection; 1988.8.3273

Feather Bonnet, White Man Runs Him, Absaroke (Crow), c. 1900 (page 98)
Eagle feathers, glass beads, ermine, porcupine quills, horsehair; 24" x 17"
Buffalo Bill Historical Center, Cody, WY
The Crow Indian Collection of Dr. William and Anna Petzolt
Gift of Genevieve Fitzgerald Estate; NA.203.934

Sitting Bull's Knife and Knife Sheath, Lakota (Sioux), Northern Plains, c. 1834
Steel, Ivory, 11.375" x 1.75"
Rawhide, tin cones, glass beads, 9.75" x 2.625"
Buffalo Bill Historical Center, Cody, WY
Dr. robert L. Anderson Collection; NA.102.91 A/B

The Buffalo People
by Emma I. Hansen, Curator, Plains Indian Museum

The buffalo, our brother, was always here with us, furnishing us food, hides for our clothes, robes for our beds, sinew, bones, everything that they provided for our livelihood. So we have a special relationship historically and religiously with the buffalo that is still strong to us this very day.

Joe Medicine Crow, Absaroke (Crow), 2000

In this way, tribal historian Joe Medicine Crow recounted the importance of the buffalo—the center of economic and spiritual life for Native people of the Great Plains.

According to Lakota tradition, the White Buffalo Woman brought the sacred buffalo calf pipe and the buffalo hunting way of life to the people. In the story, the White Buffalo Woman first appears as a *wakan* (holy) young woman but also represents the buffalo, which gave its flesh in order that the people might live. She teaches the people how to use the pipe through prayer and tells the women, "You are from the mother earth. What you are doing is as great as what the warriors do." In 1967, Lame Deer related the end of the story of the White Buffalo Woman in the following way:

The White Buffalo Woman disappeared over the horizon. Sometime she might come back. As soon as she had vanished, buffalo in great herds appeared, allowing themselves to be killed so that the people might survive. And from that day on, our relations, the buffalo, furnished the people with everything they needed—meat for their food, skin for their clothes and tipis, bones for their many tools.

For Plains Indian people, their free, buffalo hunting way of life came to an end with the adversities of the 19th century brought about through Euro-American settlement—railroads, forts, warfare, loss of lands, diseases, starvation, missionaries, government agents, and, finally by 1880, the near annihilation of the once vast herds of buffalo by commercial hide hunters. In the 1960s, Kiowa elder Old Lady Horse told the story of the destruction of the buffalo:

Then the white men hired hunters to do nothing but kill the buffalo. Up and down the plains, those men ranged, shooting sometimes as many as a hundred buffalo a day. Behind them came the skinners with their wagons. They piled the hides and bones into the wagons until they were full, and then took their loads to the new railroad stations that were being built, to be shipped east to the market. Sometimes there would be a pile of bones as high as a man, stretching a mile along the railroad track.

By the end of the 19th century, it seemed that the people themselves would disappear along with their brothers the buffalo. Remarkably, Native people of the Plains have made valiant efforts to preserve and protect traditions, ceremonies, and communities despite the poor conditions of the reservations.

In the first decade of the 20th century, the buffalo returned to the Plains through the efforts of national parks, conservationists, and ranchers in the United States and Canada. Today, as herds are established on reservations, the buffalo are again a part of daily economic and spiritual life. For Native people, this magnificent creature serves as a powerful symbol and reminder of their heritage as free hunters of the Great Plains.

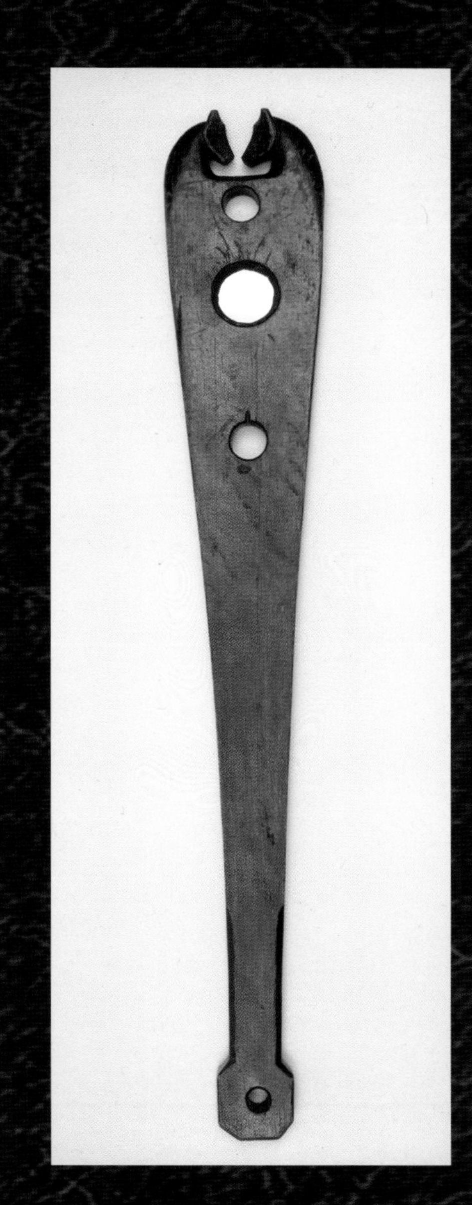

Oto dance stick, Oklahoma, c. 1900
Wood, glass mirror; 25" x. 3.75"
Buffalo Bill Historical Center, Cody, WY
Chandler-Pohrt Collection, Gift of Mr. William D. Weiss; NA.203.373

Dance sticks with carved images of horses, human heads, or other figures were carried by Plains warriors in Victory and Grass Dances to remind them of their abilities in capturing horses from enemy tribes or other battle exploits. The mirror projects a spiritual power to confuse the enemy by reflecting light into his eyes.

Crow Chief
George Catlin (1796–1872
c. 1850, oil on paper, 15.875" x 21.625"
Buffalo Bill Historical Center, Cody, WY
Bequest of Joseph M. Roebling; 7.80

George Catlin assured his many readers and those who visited his Indian Gallery that of the Indians he met during his 1832 journey up the Missouri River, the Crow were the most beautifully adorned. They were also known for their extraordinary feats of horsemanship. In *Crow Chief*, Catlin borrowed a pose from equestrian portraiture to ennoble this figure.

Indian and Pronghorn Antelope
Paul Manship (1885–1966)
1914, bronze, 13.5" x 13.5" x 8.375"; 12.5" x 8.375" x 10.5"
Buffalo Bill Historical Center, Cody, WY
William E. Weiss Fund and Mr. and Mrs. Richard J. Schwartz; 3.89A/B

Manship, a student of Solon Borglum, united a classical concern for precisely modeled forms with a modern streamlined sensibility, creating a style that was a forerunner of Art Deco. Interested in mythology, he portrayed the American Indian, not with anthropological specificity, but as a symbolic figure evoking the power of the hunt.

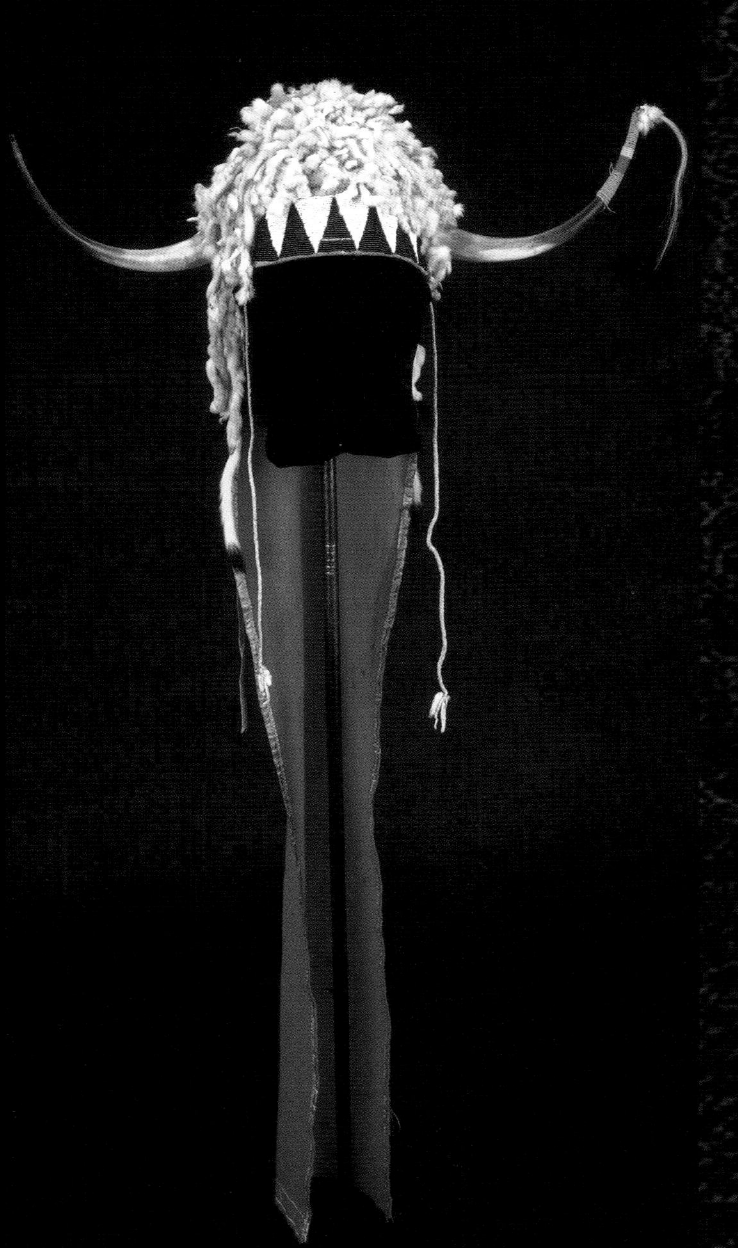

Northern Plains horn bonnet, c. 1890
Split buffalo horns, ermine skin, glass beads, wool cloth, horsehair
Buffalo Bill Historical Center, Cody, WY
Gift of Mr. And Mrs. William Henry Harrison; NA.203.19

Native people of the Plains respected and honored the buffalo—sustainer of life—through songs, dances, and ceremonies. Buffalo communicated with the people through dreams or visions and were called upon during hunger, war, illness, and other times of need. Buffalo horn bonnets manifested the animal's spiritual significance.

Meskwaki (Fox) moccasins, Tama, Iowa, c. 1880
Deer hide, glass beads; 10.25"
Buffalo Bill Historical Center, Cody, WY
Chandler-Pohrt Collection, Gift of the Pilot Foundation; NA.202.448

Thousands of tribal people were forced to move west of the Mississippi as a result of the Indian Removal Act of 1830. On reservations in Oklahoma, Nebraska, Kansas, and Iowa, they came in contact with Native people of the region. One outcome of this contact and exchange was the creation of new beadwork designs—the Prairie style—as shown in these moccasins.

A Surround of Buffalo by Indians
Alfred Jacob Miller (1810–1874)
1848–1858, o/c, 30.375" x 44.125"
Buffalo Bill Historical Center, Cody, WY
Gift of William E. Weiss; 2.76

Among the events Miller observed during his 1837 journey to the Rocky Mountains with Captain William Drummond Stewart was the buffalo hunt. As an outsider looking on, Miller described the scene: "[The Indians] all start at once with frightful yells and commence racing around the herd, drawing their circle closer and closer, until the whole body is huddled together in confusion. Now they begin firing, and as this throws them [the buffalo] into a headlong panic and furious rage, each man selects his animal."

Nakoda (Assiniboine) or ***A'anin (Gros Ventre) feather bonnet*** (page 108)
Fort Belknap Reservation, Montana, c. 1885
Eagle feathers, wool, felt, glass beads, ermine skin, horsehair; 59.5" x 18.25"
Buffalo Bill Historical Center, Cody, WY
Chandler-Pohrt Collection, Gift of Mr. And Mrs. Richard A. Pohrt; NA.203.347

Plains Indian people consider the eagle to be the most powerful of birds, and its feathers arrayed on bonnets offered protection in battle. Men earned the right to wear the feathers through courageous deeds and valor.

Nuxbaaga (Hidatsa) Man's Leggings, Northern Plains, c. 1885 (page 109)
Wool cloth, porcupine quills with pigment, glass beads, metal tacks; 32" x 9"
Buffalo Bill Historical Center, Cody, WY
Catherine Bradford McClellan Collection, gift of The Coe Foundation; NA.202.47

The daily tasks of quillworking and beadworking were vehicles of artistic expression that enriched the everyday life of the family. A well-dressed family and beautifully-crafted home furnishings demonstrated a woman's pride and love, as well as her aptitude for economic industry.

Absaroke (Crow) Shield Cover, Northern Plains, c. 1860 (page 110)
Tanned deer hide, pigment, glass beads, feathers; 20"
Buffalo Bill Historical Center, Cody, WY
Adolf Spohr Collection, Gift of Larry Sheerin; NA.108.15

The image of the grizzly bear on this shield cover symbolizes the bear's inherent strength. Other animals that appear on shields—buffalo, elk, wolves and eagles—have their own symbolic meanings.

Arikara Shield Cover, Northern Plains, c. 1875 (page 111)
Cotton cloth, pigment, feathers; 20"
Buffalo Bill Historical Center, Cody, WY
Adolf Spohr Collection, Gift of Larry Sheerin; NA.108.13

Spiritual beings appeared to men in visions and directed them to make shields and shield covers and decorate them with images that provided protection in battle.

Lakota (Sioux) Storage Bag, Northern Plains, 1890
Deer hide, dyed porcupine quills, tin cones, dyed horsehair; 14.375" x 24.25"
Buffalo Bill Historical Center, Cody, WY
Chandler-Pohrt Collection, Gift of Mr. William D. Weiss; NA.106.245

During the late 1800s to early 1900s, after the Plains tribes had been confined to reservations, an ironic flourishing of tribal arts occurred. Porcupine quillwork, which preceded beadwork, continued to be used sometimes in conjunction with other materials. At this time, women also began to illustrate men's war deeds through pictographic beading and quillwork.

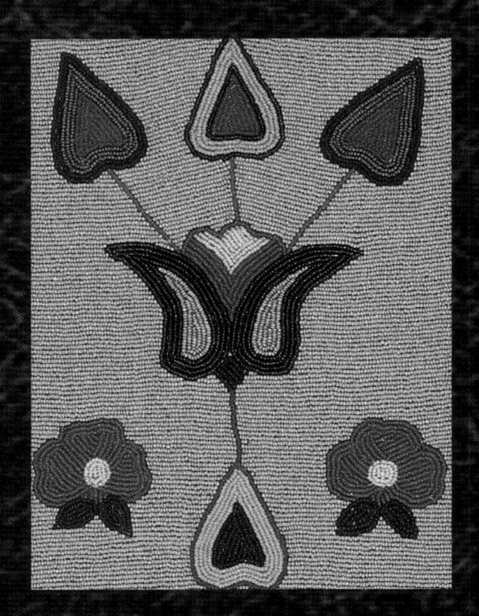

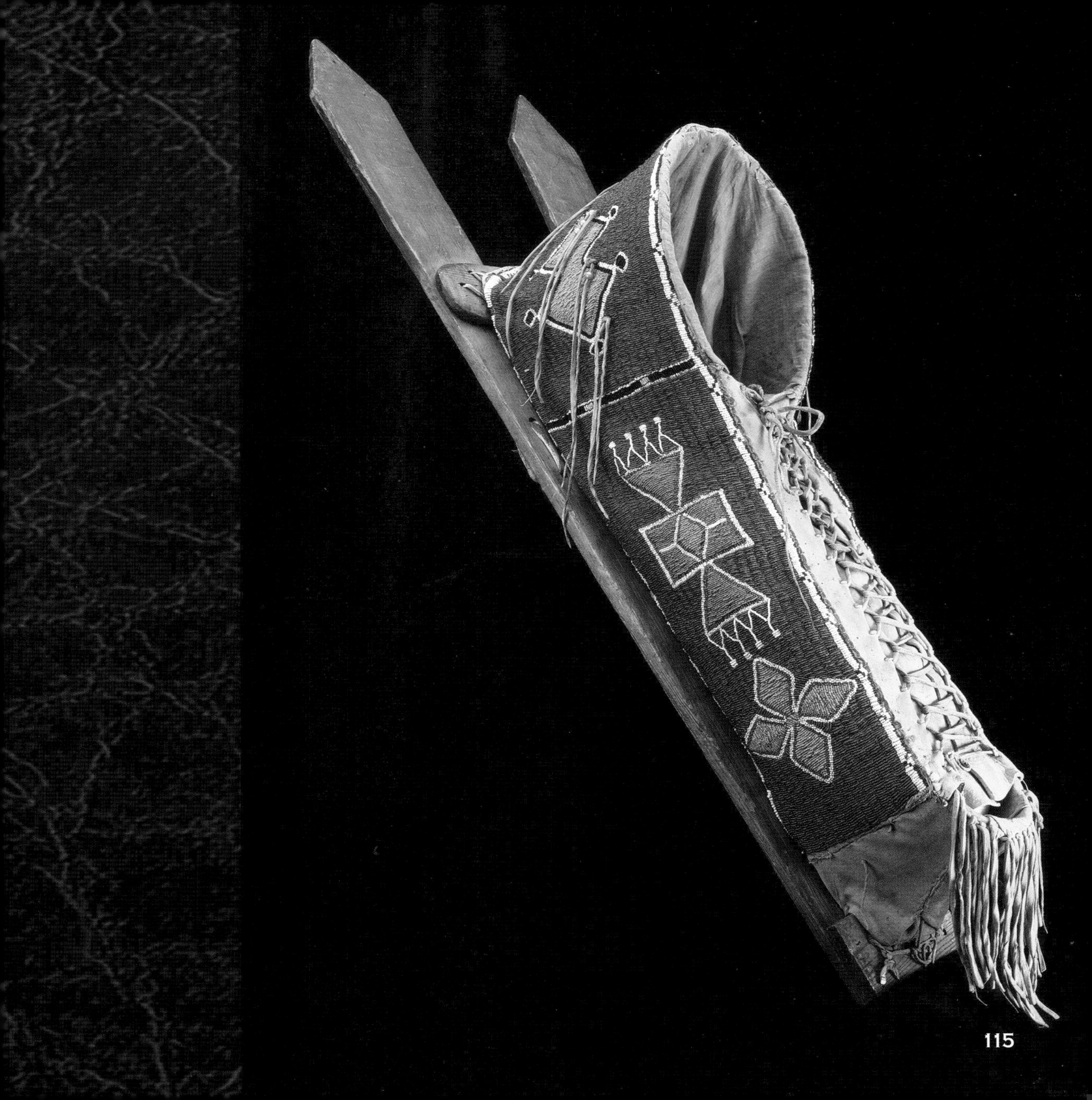

Absaroke (Crow) Cradle, Northern Plains, c. 1915 (page 114) detail (page 114)
Deer hide, wood, wool and cotton cloth, glass beads; 41.75"
Buffalo Bill Historical Center, Cody, WY
Gift of Mr. and Mrs. I.H. "Larry" Larom; NA.111.5

Among Native people, the decoration of cradles symbolizes the importance of family and children. A grandmother or another older respected woman guided the gathering of materials, design, and construction of the cradle.

Koigwu (Kiowa) Cradle, Oklahoma, 1885 (page 115)
Hide, wood, glass beads, cotton cloth; 48.25" x 12"
Buffalo Bill Historical Center, Cody, WY
Chandler-Pohrt Collection, Gift of Mr. William D. Weiss; NA.111.36

Cradles provided safe places for babies when families traveled or mothers worked. Distinctive designs were adapted to tribal lifestyles and contain important cultural symbols. For many tribes, the creation of cradles symbolized the joining of two families, with members of each taking part.

Nez Percé Babe
Edward S. Curtis (1868–1952)
Photogravure from Curtis photograph copyright 1900
Portfolio volume 8, plate 266, from *The North American Indian*, set #329/500
Cambridge, Mass.: University Press, 1907–1930
Buffalo Bill Historical Center, Cody, WY
Gift of Douglas L. Manship, Sr.

Plate 266

From Copyright Photograph 1900 by E. S. Curtis

Photogravure John Andrew & Son

KOIGWU (KIOWA) DRESS YOKE, SOUTHERN PLAINS, C. 1930
TANNED DEER HIDE, GLASS BEADS, ELK IVORIES, CLOTH; 22.75"
BUFFALO BILL HISTORICAL CENTER, CODY, WY
GIFT OF WILLIAM D. OWSLEY; NA.202.81

This dress yoke shows traditional Southern Plains design featuring pigments and long fringe. It is decorated with rows of elk teeth and detailed beaded designs and trim. Such dress yokes were worn with soft, fringed skirts made of tanned deer hide.

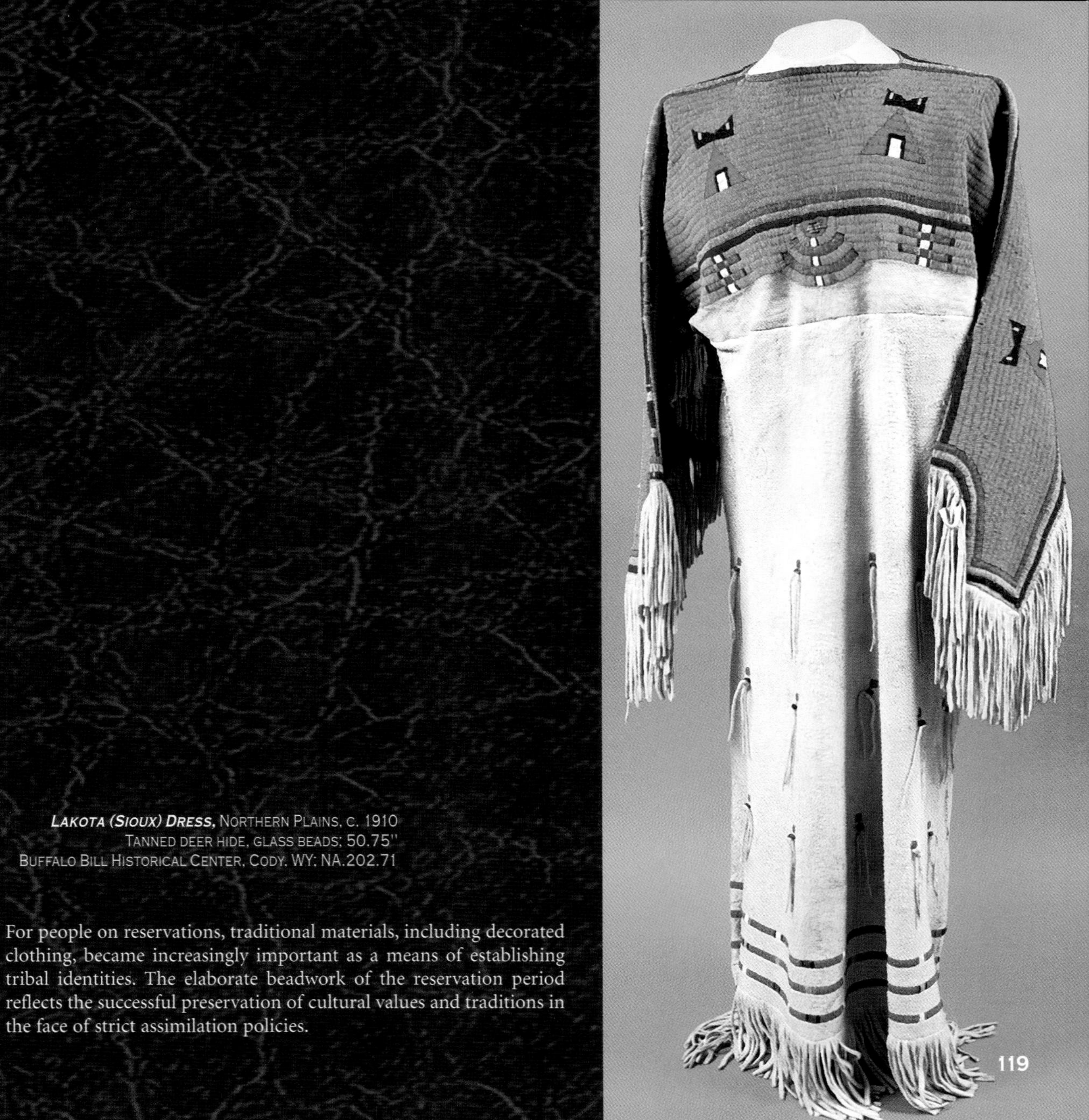

Lakota (Sioux) Dress, Northern Plains, c. 1910
Tanned deer hide, glass beads; 50.75"
Buffalo Bill Historical Center, Cody, WY; NA.202.71

For people on reservations, traditional materials, including decorated clothing, became increasingly important as a means of establishing tribal identities. The elaborate beadwork of the reservation period reflects the successful preservation of cultural values and traditions in the face of strict assimilation policies.

AKOTA (SIOUX) MAN'S VEST, NORTHERN PLAINS, C. 1880
TANNED DEER HIDE, GLASS BEADS; 19.25" X 18.5"
BUFFALO BILL HISTORICAL CENTER, CODY, WY
GIFT OF MR. AND MRS. I.H. "LARRY" LAROM; NA.202.301

The Lakota excelled in producing distinctive fully beaded clothing, moccasins and other items, often with images of warriors and horses, cowboys, deer, elk, buffalo and other animals.

TSISTSISTAS (CHEYENNE) PARFLECHE CASE,
NORTHERN PLAINS, 1885
RAWHIDE, PIGMENTS; 23.5" X 14.75"
BUFFALO BILL HISTORICAL CENTER, CODY, WY
CHANDLER-POHRT COLLECTION, GIFT OF
MR. WILLIAM D. WEISS; NA.106.147

Parfleches are folded, envelope-shaped containers fashioned from buffalo rawhide. Their geometric designs were painted using natural pigments or commercial pigments obtained through trade with Euro-Americans.

Sioux Chiefs
Edward S. Curtis (1868–1952)
Photogravure from Curtis photograph copyright 1905
Portfolio volume 3, plate 79, from *The North American Indian*, set #329/500. Cambridge, Mass.: University Press, 1907–1930
Buffalo Bill Historical Center, Cody, WY
Gift of Douglas L. Manship, Sr.

Curtis' caption for this picture is, "Very often two or three men would form themselves into a war-party and ride away to be gone weeks or months. Sometimes they returned with scalps, or horses, or women; and again the war-party, whether large or small, met defeat and none survived to bring back to anxious wives and children the story of the disaster."

A Painted Tipi–Assiniboin
Edward S. Curtis (1868–1952)
Photogravure from Curtis photograph copyright 1926
Portfolio volume 18, plate 633, from *The North American Indian*, set #329/500. Cambridge, Mass.: University Press, 1907–1930
Buffalo Bill Historical Center, Cody, WY
Gift of Douglas L. Manship, Sr.

Curtis' caption for this picture is, "A tipi painted with figures commemorative of a dream experienced by its owner is a venerated object. Its occupants enjoy good fortune, and there is no difficulty in finding a purchaser when after a few years

Buffalo Hunt (page 124)
Edgar S. Paxson (1852–1919)
1905, o/c, 26" x 38"
Buffalo Bill Historical Center, Cody, WY
Gift of Mr. & Mrs. Ernest J. Goppert, Sr., in memory of Mary Jester Allen; 86.60

In this painting, Paxson offers an unusual perspective of the buffalo hunt. Rather than passive and docile, the buffalo turns and aggressively charges the hunter, startling both him and his horse. Paxson referred to the work by differing titles. In a June 23, 1905 diary entry, Paxson wrote, "I began working on a new oil, *Turning the Worm*, a buffalo hunt where the buffalo has turned on the hunter." A November 24, 1905 diary entry indicates Paxson had finished the painting, calling it *When He is Bad.*

Sharps M1869 Military Rifle, Adapted for Buffalo Hunting by Freund & Bro. (page 124)
Maker: Sharps Rifle Company, Hartford, CT and Frank W. Freund & Bro.; Cheyenne, Wyoming Territory
Serial Number: 155,418
Caliber: .50 (.50-90 centerfire cartridge)
Barrel Length: 30 inch (bore)
Date: About 1875 (original production) and 1877–1881 (adaptations)
Buffalo Bill Historical Center, Cody, WY; 1983.7.1

Henry Gerdel of Big Horn Valley, Wyoming used this rifle in the buffalo trade.

Buffalo Coat, c. 1890 (page 125)
Buffalo hide and fur, shorn sheepskin, satin
and cotton cloth, celluloid buttons
Buffalo Bill Historical Center, Cody, WY
Gift of Dave Jones; 1.69.1242

Commercial hide hunters began exterminating buffalo in the 1860s, with the help of the U.S. military. By 1883, they had destroyed the last wild herds.

After the Buffalo Run, North Montana, 1879
L. A. Huffman (1854–1931)
Photographic print from an 1879
glass negative, hand colored
Buffalo Bill Historical Center, Cody, WY
Gift of Thomas Minckler; MS-100 LAH 1979

Huffman arrived in Montana in time to witness and document the last of the free-ranging buffalo herds on the northern plains and the last of the hide hunters still at work.

After The Buffalo Run. North Mont 1879.
© L A Huffman.

Exhibition Artwork Index

John James Audubon (1785–1851)
Northern Hare, Winter Pelage
lithograph Pg. 24

John James Audubon (1785–1851)
Northern Hare (old & young), Summer Pelage
lithograph Pg. 25

John James Audubon, artist (1785–1851)
Robert Havell, Jr., engraver (1793–1878)
The Bird of Washington
hand colored, 37.75" x 24.75" Pg. 23

Irving R. Bacon (1875–1962)
The Conquest of the Prairie
o/c, 47.25" x 118.5" Pg. 78–79

James Bama (b. 1926)
A Contemporary Sioux Indian
oil on panel, 23.375" x 35.375" Pg. 86

Albert Bierstadt (1830–1902)
Island Lake, Wind River Range, Wyoming
o/c, 26.5" x 40.5" Pg. 9

Rosa Bonheur (1822–1899)
Col. William F. Cody
o/c, 18.5" x 15.25" Pg. vi & 35

George Catlin (1796–1872)
Crow Chief
oil on paper, 15.875" x 21.625" Pg. 102

Cyrus E. Dallin (1861–1943)
Appeal to the Great Spirit
bronze, 21" x 21.5" x 15" Pg. 89

Douglas Duer (1887–1964)
"A Woman! A Girl! I've Killed a Girl!"
o/c, 23" x 16.5" Pg. 91

Harvey Dunn (1884–1952)
The Scout
o/c, 36.125" x 24.125" Pg. 82

William Herbert Dunton (1878–1936)
The Custer Fight
o/c, 32.25" x 50.25" Pg. 40–41

Henry Farny (1847–1916)
Days of Long Ago
o/b, 37.5" x 23.75" Pg. 2

James Earle Fraser (1876–1953)
End of the Trail
bronze, 33.75" x 26" x 8" Pg. 93

Philip R. Goodwin (1881–1935)
Trapper Attacked by Wolves
o/c, 29.25" x 15" Pg. 12

William Jacob Hays, Sr. (1830–1875)
A Herd of Bison Crossing the Missouri River
o/c, 36.125" x 72" Pg. 8

Malvina Hoffman (1887–1966)
Sign Talker-Blackfoot Man
bronze, 28.5" x 8.25" x 6.5" Pg. 87

Wilson Hurley (b. 1924)
View from the Mohave Wall
o/c, 60.25" x 90.25" Pg. 18–19

Henry Inman (1801–1846)
Shar-I-Tar-Ish
o/c, 30.375" x 25.625" Pg. 45

Harry Jackson (b.1924)
Indian Mother and Child
bronze, painted, 27.75" x 35.25" x 29.75" Pg. 36

W.H.D. Koerner (1878–1938)
Madonna of the Prairie
o/c, 37" x 28.75" Pg. 48

William R. Leigh (1866–1955)
Panning Gold, Wyoming
o/c, 32.25" x 40" Pg. 17

James Otto Lewis (1799–1858)
Waa-Pa-Laa or The Playing Fox, in *The Aboriginal Portfolio* Pg. 56

Henry Lion (1900–1966)
Lewis and Clark and Sacagawea
bronze, 35.25" x 23.75" x 28.25" Pg. viii

Paul Manship (1885–1966)
Indian and Pronghorn Antelope
bronze, 13.5" x 13.5" x 8.375";
12.5" x 8.375" x 10.5" Pg. 103

Thomas L. McKenney (1785–1859) and
James Hall (1793–1868)
Mo-Hon-Go an Osage Woman, in *History of the Indian Tribes of North America* Pg. 57

Alfred Jacob Miller (1810–1874)
A Surround of Buffalo by Indians
o/c, 30.375" x 44.125" Pg. 106–107

Thomas Moran (1837–1926)
Golden Gate, Yellowstone National Park
o/c, 36.25" x 50.25" Pg. 10–11

William Notman and Son
Sitting Bull cabinet card Pg. 63

Edgar S. Paxson (1852–1919)
Custer's Last Stand
o/c, 70.5" x 106" Pg. ii–iii & 38–39

Edgar S. Paxson (1852–1919)
Buffalo Hunt
o/c, 26" x 38" Pg. 124

Alexander Pope (1849–1924)
Weapons of War
o/c, 54" x 42.5" x 2" Pg. 37

William Tylee Ranney (1813–1857)
The Prairie Burial
o/c, 28.5" x 41" Pg. 49

Frederic Remington (1861–1909)
The Rattlesnake
bronze, 22.625" x 22.625" x 13" Pg. 15

Frederic Remington (1861–1909)
The Broncho Buster
bronze, 23.375" Pg. 77

Frederic Remington (1861–1909)
The Buffalo Hunt
o/c, 34" x 49" Pg. 88

Carl Rungius (1869–1959)
Mule Deer in the Badlands, Dawson County, Montana
o/c, 59.625" x 75.25" Pg. 16

Charles M. Russell (1864–1926)
Nature's Cattle
bronze, 4.75" x 4" x 15.325" Pg. 27

Charles M. Russell (1864–1926)
Waiting for a Chinook
w/c on paper, 20.5" x 29" Pg. 42

Charles M. Russell (1864–1926)
Self Portrait
w/c on paper, 12.375" x 6.875" Pg. 80

Charles M. Russell (1864–1926)
The Bucker and the Buckeroo
bronze, 14.75" x 8" x 10.5" Pg. 90

Samuel Seymour (c. 1775–1823)
Distant View of the Rocky Mountains
(frontispiece), 4.437" x 7.25" Pg. 5

Bill Schenck (b. 1947)
A Flight from Destiny
o/c, 47.375" x 62.375" Pg. 20

Fritz Scholder (b. 1937)
Aspen Summer
o/c, 40.25" x 30.125" Pg. 21

Charles Schreyvogel (1861–1912)
The Summit Springs Rescue, 1869
o/c, 48" x 66" Pg. 46–47

Joseph Henry Sharp (1859–1953)
The Broken Bow
o/c, 44.5" x 59.375" Pg. 83

Henry Merwin Shrady (1871–1922)
Buffalo
bronze, 12.25" x 15.5" x 6.5" Pg. 26

Sitting Bull (1831–1890)
Sitting Bull Killing a Crow Indian
pencil and crayon on paper, 8.5" x 10.25"
Pg. 81

John Mix Stanley (1814–1872)
Last of Their Race
o/c, 43" x 60" Pg. 6–7

Stacy
Annie Oakley signed cabinet card Pg. 62

N.C. Wyeth (1882–1945)
Cutting Out
o/c, 38" x 25.875" Pg. 84

Exhibition Aritifacts Index

Absaroke (Crow) Shield Cover
Tanned deer hide, pigment, glass beads, feathers; 20" Pg. 110

Absaroke (Crow) Cradle
Deer hide, wood, wool and cotton cloth, glass beads; 41.75" Pg. 114

Annie Oakley's glove
Leather with brass snap Pg. 60

Arikara Shield Cover
Cotton cloth, pigment, feathers; 20" Pg. 111

Bridle and martingale
Black leather with cowrie shells and brass hardware Pg. 66

Buffalo Bill's beaded buckskin jacket
Tanned buckskin, glass beads, brass buttons, blue satin Pg. 68

Buffalo Coat
Buffalo hide and fur, shorn sheepskin, satin and cotton cloth, celluloid buttons Pg. 125

Exploratory Travels through the Western Territories of North America...
Capt. Zebulon Montgomery Pike (1779–1813)
Pg. 58–59

Feather Bonnet, White Man Runs Him, Absaroke (Crow)
Eagle feathers, glass beads, ermine, porcupine quills, horsehair; 24" x 17" Pg. 98

Hinono'ei (Southern Arapahoe) Ghost Dance shirt
Elk hide, eagle feathers, pigments, 40" x 29" Pg. 28

Holograph letter to Gen. George Crook
Gen. Alfred H. Terry (1827–1890) Pg. 55

Koigwu (Kiowa) Cradle
Hide, wood, glass beads, cotton cloth; 48.25" x 12" Pg. 115

Koigwu (Kiowa) Dress Yoke
Tanned deer hide, glass beads, elk ivories, cloth; 22.75" Pg. 118

Lakota (Sioux) Man's Vest
Tanned deer hide, glass beads; 19.25" x 18.5" Pg. 120

Lakota (Sioux) Dress
Tanned deer hide, glass beads; 50.75" Pg. 119

Lakota (Sioux) dress
deer hide, glass beads, 63" x 58" Pg. 29

Lakota (Sioux) Storage Bag
Deer hide, dyed porcupine quills, tin cones, dyed horsehair; 14.375" x 24.25" Pg. 113

Life on the Plains, Or Personal Experiences with Indians, 1874
Gen. George A. Custer (1839–1876) Pg. 53

Meskwaki (Fox) moccasins
Deer hide, glass beads; 10.25" Pg. 105

Mission Saddle
Edward H. Bohlin
Brown leather, silver Pg. x

Nakoda (Assiniboine) or A'anin (GrosVentre) feather bonnet
Eagle feathers, wool, felt, glass beads, ermine skin, horsehair; 59.5" x 18.25" Pg. 108

Northern Plains horn bonnet
Split buffalo horns, ermine skin, glass beads, wool cloth, horse hair Pg. 104

Nuxbaaga (Hidatsa) Man's Leggings
Wool cloth, porcupine quills with pigment, glass beads, metal tacks; 32" x 9" Pg. 109

Oglala Lakota (Sioux) shirt
Tanned deer hide, glass beads, human hair, ermine, wool cloth, feathers, porcupine quills with pigment; 37" Pg. 51

Oto dance stick
Wood, glass mirror; 25" x 3.75" Pg. 101

Peace medal, 1849
Bronze, brass beads; 3" Pg. 44

Remington rifle pin
10K gold Pg. 60

Sitting Bull's Knife and Knife Sheath, Lakota (Sioux)
Steel, Ivory, 11.375" x 1.75"
Rawhide, tin cones, glass beads, 9.75" x 2.625" Pg. 99

Tsistsistas (Cheyenne) Parfleche Case
Rawhide, pigments; 23.5" x 14.75" Pg. 121

Über die Selbstständigkeit der Species des Ursus Ferox, Desm. 1856
Prinz Max von Weid (1782–1867) and Dr. C. Mayer Pg. 52

Wildfire, book
The Mysterious Rider, book Pg. 95
Zane Grey

Exhibition Firearms Index

Colt's Patent Firearms Mfg. Co., Hartford, CT
Colt Model 1851
Percussion Revolvers Pg. 96

Colt's Patent Firearms Mfg. Co., Hartford, CT
Engraved Colt Model 1849 "Pocket" 4.5" Barrel
Percussion Revolver Pg. 97

Colt's Patent Firearms Mfg. Co., Hartford, CT
Colt Model 1873 "Frontier" Single Action Army
Revolver Engraved Pg. 97

Harpers Ferry Armory, Harpers Ferry, VA
U.S. M1803 Flintlock Rifle Pg. 72

Jacob and Samuel Hawken
Hawken .56 caliber half stock percussion rifle
Pg. 72

Isaiah Lukens; c. 1800–1810
Lukens (Butt Reservoir) Air Rifle Pg. 73

E. Remington & Sons
Buffalo Bill's "No. 1" Military Rifle in .43
Spanish Pg. 69

Sharps Rifle Company & Frank W. Freund & Bro.
Sharps M1869 Military Rifle, Adapted for
Buffalo Hunting by Freund & Bro. Pg. 124

Springfield Armory
U.S. M1873 Carbine Pg. 73

Winchester Repeating Arms Co., New Haven, CT
Winchester Model 1907 Deluxe Semi-Automatic
Rifle Pg. 4

Winchester Repeating Arms Co., New Haven, CT
Winchester Model 1886 Lever Action Rifle
Pg. 14

Winchester Repeating Arms Co., New Haven, CT
Winchester Model 1895 Lever Action Rifle
Pg. 14

Winchester Repeating Arms Co., New Haven, CT
Annie Oakley's Model 92 carbine in .32 WCF
Pg. 60

Winchester Repeating Arms Co., New Haven, CT
Frederic Remington's Winchester M1894 Deluxe
Sporting Rifle Pg. 96

Winchester Repeating Arms Co., New Haven, CT
Zane Grey's Winchester M1895
Sporting Rifle Pg. 95

Exhibition Photo Index

Charles J. Belden
A Long, Long Trail A-Winding
Photograph Pg. 43

Buffalo Bill's Wild West Show Personnel, c. 1912
Photograph, 11" x 64.375" Pg. 74–75

Edward S. Curtis (1868–1952)
White Man Runs Him-Apsaroke
Photogravure from Curtis photograph Pg. 92

Edward S. Curtis (1868–1952)
Nez Percé Babe
Photogravure from Curtis photograph Pg. 117

Edward S. Curtis (1868–1952)
Sioux Chiefs
Photogravure from Curtis photograph Pg. 122

Edward S. Curtis (1868–1952)
A Painted Tipi–Assiniboin
Photogravure from Curtis photograph Pg. 123

L. A. Huffman (1854–1931)
After the Buffalo Run, North Montana 1879
Photographic print Pg. 127

L. A. Huffman (1854–1931)
A Hot Noon beside the Round-up Camp, Big Dry, Montana
1879 glass neg. hand colored print Pg. 32–33

Shooting over Gibson Girl the 2nd
Photographic print Pg. 64–65

Exhibition Poster Index

Works included in the catalogue are indicated by page number

A. Hoen & Co.
A Colony of Genuine Mexican Vaqueros
Color lithograph poster, 20.25" x 28.375"
Pg. 31

A. Hoen & Co.
Miss Annie Oakley, The Peerless Lady Wing-shot
Color lithograph poster Pg. 61

Courier Co.
Custer's Last Stand
Color lithograph poster Pg. 70–71

Forbes' Lithographic Co.
A Bucking Mustang
Color lithograph poster, 42" x 29"

Forbes' Lithographic Co.
Steer Riding
Color lithograph poster, 42" x 29"

Philip R. Goodwin (1881–1935)
Trapper Attacked by Wolves
chromolithograph, 27.443" x 15.313" Pg. 13

Strobridge Lithographic Co.
Portrait
Color lithograph poster, 40" x 58"

The Enquirer Job Printing Co.
Poster, 1898
color lithograph poster, 110.5" x 81" Pg. xii

U.S. Lithograph Co.
"The Farewell Shot" Positively the last appearance of Col. W.F. Cody (in the saddle), "Buffalo Bill."
Color lithograph poster, 28" x 41" Pg. 67